Bringing Society Back In

American and Comparative Environmental Policy
Sheldon Kamieniecki and Michael E. Kraft, editors

Bringing Society Back In

Grassroots Ecosystem Management,
Accountability, and Sustainable Communities

Edward P. Weber

The MIT Press
Cambridge, Massachusetts
London, England

This book was set in Sabon by Achorn Graphic Services, Inc., with the Miles 33 system.

Printed and bound in the United States of America.

Library of Congress Cataloging-in-Publication Data

Weber, Edward P.
  Bringing society back in : grassroots ecosystem management, accountability, and sustainable communities / Edward P. Weber.
      p.   cm. — (American and comparative environmental policy)
  Includes bibliographical references (p.).
  ISBN 0-262-23226-X (hc. : alk. paper) — ISBN 0-262-73151-7 (pbk. : alk. paper)
  1. Environmental policy—Citizen participation. 2. Environmental policy—Northwest, Pacific—Citizen participation—Case studies. 3. Environmental protection—Citizen participation. 4. Environmental protection—Northwest, Pacific—Citizen participation—Case studies. I. Title. II. Series.

GE180 .W425 2003
363.7′0525—dc21

                                                                2002032163

10 9 8 7 6 5 4 3 2 1

For Natalie Meyer,
a mentor extraordinaire,
thank you for pushing me and reminding me of my potential over the years. Your keen insight into politics and the human condition, and your willingness to share it with me, have been much appreciated.

Contents

Foreword

Some of the most bitter controversies over U.S. environmental policies have occurred in small Western communities where timber, mining, and ranching interests have clashed with those seeking to preserve public lands for ecological or recreational purposes. Whether the conflict was over implementation of the Endangered Species Act, the National Forest Management Act, or any number of other federal or state laws, the outcome often was policy stalemate and local economic stagnation. At the national level, the experience in these cases stimulated ideological debates over the relative importance of economic development and environmental protection, and fed an antienvironmental movement that sought to weaken federal statutes thought to contribute to these conflicts.

Against that background, one of the most intriguing developments of the past decade has been the rise of grassroots governance efforts in such Western communities, and elsewhere around the nation. Those directing these efforts have sought to reconcile competing values through collaborative and participatory decision making that brings together citizens, key stakeholder groups, and government agencies in a search for acceptable solutions. These ad hoc and voluntary processes have helped to foster consensus on habitat conservation plans for protecting endangered species, restoration efforts for degraded ecosystems, smart-growth strategies for suburban communities, and redevelopment of contaminated lands. These experiments highlight the importance of inquiry into how such grassroots environmental decision making actually works, how well it meets expectations for political accountability, how successful it is in achieving desired environmental outcomes, and the conditions that contribute to its success over time.

In this book Edward P. Weber uses case studies in what he calls *grassroots ecosystem management* (GREM) to explore these and other related questions. This movement is found mainly in rural communities in the West, where local economies have been tied closely to natural resource use, such as timber harvesting. He examines three cases that appear typical of successful efforts at grassroots governance, or "bringing society back in": Willapa Bay, Washington, the Henry's Fork watershed in Idaho, and the Applegate Valley in Oregon. He offers a detailed description of these experiments from the perspective of the participants, based on extensive personal interviews as well as the documentary record.

Critics of GREM have expressed concern over whether extractive-industry interests would dominate local decision making, and thus contribute to further environmental degradation. If that were indeed the result, it would be hard to defend these new community arrangements or recommend their use over established agency procedures. Thus, Weber asks whether political accountability is possible when decentralized, collaborative, and participatory institutions are relied on in this way. His study also examines the relationship between accountability and environmental policy performance. That is, to what extent do decentralized and collaborative efforts of this kind actually improve environmental conditions? Weber finds that GREM can be simultaneously accountable to a diversity of individuals, communities, surrounding regions, and the nation. He argues that the process can produce win-win outcomes and help to integrate environmental and economic values. These efforts can focus a community's attention on environmental sustainability while also building its institutional capacity to ensure the kind of future collectively desired by residents.

Weber's work illustrates the kind of books published in the MIT Press series in American and Comparative Environmental Policy. We encourage work that examines a broad range of environmental policy issues. We are particularly interested in volumes that incorporate interdisciplinary research and focus on the linkages between public policy and environmental problems and issues both within the United States and in cross-national settings. We welcome contributions that analyze the policy dimensions of relationships between humans and the environment from either an empirical or a theoretical perspective. At a time when environmental policies are increasingly seen as controversial and new approaches are being implemented widely, we are especially interested in studies that assess policy successes and failures, evaluate new institutional arrangements and policy tools, and clarify new directions for environmental politics and policy. The books in this series are written for a wide audience that includes academics, policymakers, environmental scientists and professionals, business and labor leaders, environmental activists, and students concerned with environmental issues. We hope they contribute to public understanding of the most important environmental problems, issues, and policies that society now faces and with which it must deal well into the new century.

Sheldon Kamieniecki, University of Southern California
Michael E. Kraft, University of Wisconsin–Green Bay
American and Comparative Environmental Policy Series editors

Preface and Acknowledgments

In the late 1980s a growing number of people across the United States started creating and choosing new paradigms for governance in a multiplicity of policy arenas, including education, policing, rural development, public health, and tax administration. In the environmental policy field, the movement toward alternative institutions involved hundreds (some estimate thousands) of decentralized, collaborative, and participative governance arrangements that rely on deliberation, consensus, and a holistic management approach. These new efforts seek simultaneously to enhance governance performance *and* accountability.

The growth of these new governance efforts, however, set off alarms within some circles, including the national environmental advocacy community. According to their analysis, the new institutions would necessarily result in special interest government, the acceleration of environmental degradation, and an end run around national environmental protection laws—a far cry from the claims for improved accountability and policy performance.

On the other hand, a burgeoning body of research suggests that the participants in these new governance arrangements might be on to something. It might be possible to improve accountability to local interests, both private and public, without diminishing accountability to broader, national interests. In addition, many scholars and practitioners now contend that effective environmental policy programs require new rules of engagement that recognize the critical importance of social complexities and that strengthen collaboration among diverse government, civic, and business actors at the state and local levels.

This book takes a first step in the direction of establishing a theoretical framework for understanding the puzzles of accountability and policy performance raised by these new governance arrangements. It does so by studying the actual practices of politics and policymaking. It explores the characteristics and operational dynamics of the various accountability mechanisms being utilized to find out how accountability works and to identify some of the conditions under which decentralized, collaborative, and participatory policy administration arrangements are most likely to achieve accountability. In addition, the book investigates the connection between accountability and policy performance, or the actual outcomes being produced by these cases. Of specific concern is how the outcomes accord with any claims for broad-based accountability and with the expectations of participants for environmental policy outcomes that are not only simultaneously supportive of the environment and economy, but also sensitive to the goal of environmental sustainability.

This book could not have been completed without the assistance of a great many people. A large number of public officials, representatives of various interests, and citizens from the rural communities that are the focus of this book gave generously of their time to answer my questions, help me locate pertinent materials, read drafts of various portions of the manuscript, and clarify the operational dynamics of their grassroots institutions as well as the different pieces of the accountability puzzle. Just as importantly, many people opened their doors to me—a stranger—and helped me to understand the forests, farms, waters, and communities of their special places. Others not only opened their doors, they helped me gain access to key participants in these new governance arrangements. Especially helpful in this regard were Janice Brown and Susan Steinman of the Henry's Fork Foundation, Dale Swenson of the Fremont-Madison Irrigation District, Su Rolle of the U.S. Forest Service (and Bureau of Land Management), Jack Shipley and Jan Perttu of the Applegate Partnership, Dan'l Markham of the Willapa Alliance, David Campiche, owner of the Shelburne Inn, and John McMahon of Weyerhaeuser.

I am also indebted to a whole host of scholars. Lawrence O'Toole, Anne Khademian, Nicholas Lovrich, and Lance LeLoup read and critiqued key parts of the manuscript during the early stages, in addition to providing useful responses to the original book prospectus. Their feedback, insight, and encouragement convinced me that this was a project worth doing, and made the book much stronger. Matthew Carroll, Riley Dunlap, Richard Krannich, and Mark Brunson gave timely and much-needed advice on the idea of grassroots ecosystem management as a new environmental movement. Co-collaborators on a number of journal articles and book chapters related to the larger research agenda on grassroots ecosystem management, including Brent Steel, Philip Brick, Bruce Shindler, and Christina Herzog, helped me to see key concepts and arguments from entirely different perspectives. Gary Wamsley, Barbara Romzek, Evan Ringquist, Charles Davis, Mark Lubell, David Nice, John Freemuth, Gregory Walker, and Steven Daniels, along with a number of conference participants and discussants, supplied helpful commentary at different stages of the project. The four anonymous outside reviewers at MIT Press challenged me to clarify and revise key segments of the original manuscript, while the series editors, Sheldon Kamieniecki and Michael Kraft, provided invaluable guidance along the way and were instrumental in strengthening the book's arguments. Christina Herzog and Michael Massoglia provided invaluable research assistance, while Michael Gaffney, John McGuire, Ellen Lemley, Ira Parnerkar, and Christina Hannum-Buffington, graduate students at Washington State University, kept me on my toes and made me a better teacher and scholar.

I am also grateful for the generous support I received from the Washington State University College of Liberal Arts and Department of Political Science. The Edward R. Meyer Fund of Washington State University provided essential field research funds. Robert Breckenridge of the Idaho National Environmental and Energy Laboratory (INEEL) supported a portion of this research by sponsoring an examination of how collaborative science worked in the case of the Henry's Fork Watershed Council. Many thanks also to William Budd and George Hinman of the Environmental Science and Regional Planning program at Washington State University for giving me the opportunity to present pieces of the research in their graduate student brown-bag forums.

Others supported this research by sponsoring its presentation at various forums around the country. The Natural Resources and Environmental Policy Center at Utah State University, under the able direction of Richard Krannich, sponsored a visit and guest lecture in fall 1998. The Henry's Fork Watershed Council invited me to give their keynote address at their 1998 annual state-of-the-watershed conference. I was also fortunate enough to participate in the first workshop of the National Consortium for Community-Based Collaboratives sponsored by the University of Arizona, the Udall Public Policy Center, and the University of Virginia (Tucson, Arizona, October 1999). And thanks to Alnoor Ebrahim of the Virginia Tech Urban Affairs and Planning Department, who, in cooperation with the School for Public and International Affairs, sponsored the first full presentation of the book's arguments in April 2001.

Nor would this book have been possible without the loving support of my wife, Andrea, and the inspiration (as well as "reality" breaks) provided by Nicholas, Cody, and Alexis—the three best kids a dad could ever hope for (and yes, one is now a teenager). Also deserving of heartfelt gratitude is Lucinda Miller, the career counselor of many years ago who patiently walked me along the path of self-discovery and convinced me to ignore the central question, "How much money will I make," that seems to drive so many career choices today. Accepting her challenge to choose a career for the love of it, rather than the money, is something I will never regret. Finally, thanks to all those in the Henry's Fork, Applegate, and Willapa areas who gave me advice on where to find fish without crowds, and what flies were most likely to tempt the fish, whether it was rainbows, cutthroats, steelhead, or coho.

Although I have benefited tremendously from countless sources of help, the views expressed in this book, as well as any shortcomings, are the sole responsibilities of the author.

1

Changing Institutions, Accountability, and Policy Performance

Quincy, California, is a small logging town of 5,000 people near Lake Tahoe in the northeastern part of the Sierra Nevada Mountains. For most of the 1980s, the community was at war. Loggers intent on extracting raw materials from timber-rich forests and environmentalists just as intent on protecting watersheds and forests from perceived ruin fought legal battles in the courts and brawled in local taverns and coffeehouses. Over time, environmentalists' legal appeals and lawsuits succeeded in virtually shutting down the local timber supply.

In 1992, three erstwhile adversaries decided that the prospect of interminable gridlock was unacceptable. They teamed up in search of peace and, perhaps even more difficult, a forest management prescription capable of simultaneously meeting the goals of environmentalists, timber interests, and local politicians interested in sustaining the local economy. Tom Nelson, the logging industry's top lobbyist in California, along with Plumas County Supervisor Bill Coates, initiated the search for a solution when they called their political archenemy, Michael Jackson. Jackson was the local environmental attorney responsible for much of the litigation stopping the timber industry from conducting business as usual.

Nelson and Coates proposed using a selective logging plan forwarded by environmentalists in 1986 as a starting point in the negotiation to reopen the forests to logging, rebuild the local economy, and protect the environment. The plan called for no more logging in old growth, no more roads in roadless areas, larger vegetative buffers in riparian zones, and selective cutting on the surrounding national forests to restore forest health and protect people from cataclysmic fires. The plan also prescribed timber-harvest levels four times higher than the current U.S. Forest Service (USFS) proposal for clear-cuts.

The group, along with other citizens, met a number of times at the local library before hosting a public meeting in the spring of 1993 at the Quincy Town Hall Theatre. Quincy Library Group (QLG) participants explained that they were trying to end the timber wars by finding common ground. When all was said and done, all but 5 of the 250 people in attendance gave their blessing to the collaborative attempt to forge a win-win scenario for forest management. Several months later the QLG published its Community Stability Proposal based on the environmentalists' 1986 plan. Of the 2.5 million acres covered by the proposal, almost 1 million acres were placed in various reserves off limits to logging, including wilderness areas. The off-limits areas included 148,000 acres of roadless, old-growth land that was open to logging under the USFS's existing forest management plan. Environmentalists also secured expanded protection for riparian zones, protected all trees over 30 inches in diameter, and ended large clear-cuts (40 acres) in favor of small openings (less than 2 acres) and selective thinning timber sales. In exchange, local communities would get jobs and fire protection, while timber companies were allowed to extract wood from 1.6 million acres of managed land at a volume roughly double that of existing USFS logging levels (compared to the original 1986 proposal of four times the volume) (Braxton 1995; Marston 1997b, 1, 8–9).

The QLG soon became famous as a model for collaborative resource management[1]: "The group enjoyed extensive media coverage and was celebrated by the Clinton administration as an example of a collaborative approach promising a win-win outcome for the ecosystem and the economy. The White House blessed the effort by choosing the nation's Christmas tree from the Plumas National Forest near Quincy" and by providing $9 million over a three-year period to implement the agreement.[2] But the USFS balked at implementing the agreement, and QLG participants decided to take their case to the nation's capital in 1997.

Political support for the QLG plan at the federal level ranged across the spectrum from "green democrats" like U.S. Representatives Peter DeFazio of Oregon and Vic Fazio of California as well as Senators Dianne Feinstein (D) and Barbara Boxer (D) of California, to Republican environmental leader Sherwood Boehlert of New York, and to California Governor Pete Wilson and other Western Republicans typically not supportive of environmental measures.[3] Legislation to implement the plan eventually passed the U.S. House of Representatives by a 429-to-1 margin (H.R. 858). And although there were similar levels of support in the Senate, national environmental advocacy groups such as the Sierra Club and Wilderness Society used their considerable clout to stymie the bill's passage. Senator Feinstein eventually ensured passage of the QLG bill by attaching it as a rider to the 1999 federal spending bill. President Clinton signed the bill into law on October 21, 1998 (Davis and King 2000).

The QLG is not alone in its efforts. Starting in the late 1980s and early 1990s a growing number of people became tired of fighting among themselves, upset with the limitations of the top-down, command-oriented, fragmented natural resource and environmental policy management regimes, and fearful of the negative effects of increased development pressures for both the environment and the character of their communities. In search of better governance performance *and* enhanced accountability to a broader array of interests, coalitions of the unalike—citizens, government regulators, small businesses, environmentalists, commodity interests, and others—are creating and choosing alternative institutions for governing public lands and natural resources.[4]

At the forefront of this movement toward alternative institutions is grassroots ecosystem management (Weber 2000a), or what others have called community-based conservation (Western and Wright 1994), watershed democracy,[5] cooperative ecosystem management (Yaffee et al. 1996), community conservation (Snow 1996), collaborative conservation (Cestero 1999), and the watershed movement (Rieke and Kenney 1997; Born and Genskow 1999).[6] By *grassroots ecosystem management* (GREM), I mean an ongoing, collaborative governance arrangement in which inclusive coalitions of the unalike come together in a deliberative format to resolve policy problems affecting the environment, economy, and community (or communities) of a particular place. Such efforts are governance arrangements because the act of governing involves

the establishment and operation of social institutions or, in other words, sets of roles, rules, decision making procedures, and programs that serve to define social practices and to guide interactions of those participating in these practices. . . . Politically significant institutions or governance systems are arrangements designed to resolve social conflicts, enhance social welfare, and, more generally, alleviate collective action problems in a world of interdependent actors. Governance, on this account, does not presuppose the need to create material entities or organizations—"governments"—to administer the social practices that arise to handle the function of governance. (Young 1996, 247; Young 1994; Ellickson 1991; North 1990)

In more specific terms, GREM organizes on the basis of geographic "place," and is intergovernmental in character (rather than being strictly federal or state based). The local "places" are rural economies dependent on nature's bounty, whether in the form of agricultural commodities, forestry products, commercial fisheries, outdoor recreation and tourism activities, or ranching.[7] The biophysical, geographic scale of "place" varies and is the product of political, rather than scientific, agreement among those involved in each effort. As such, "place" is often defined as a valley and its surrounding topography, or as a watershed, rather than as an ecosystem per se. In fact, the term *ecosystem* in ecosystem management connotes the crosscutting, holistic, comprehensive approach to the notion of conservation and management that focuses on environmental protection, economic development, and community well-being, rather than on the specific biophysical scale of management. Put differently, participants in GREM efforts seek to manage valleys, watersheds, forests, or landscapes as a whole, rather than in fragmented, piecemeal fashion.

GREM also relies extensively or exclusively on collaborative decision processes, consensus, and active citizen participation, which means that private citizens and stakeholders often take on leadership roles and are involved directly in deliberative decision-making, implementation, and enforcement processes along with government officials, especially when it comes to how goals are to be achieved. Decision making typically involves a broad-based coalition of the unalike. This means that loggers and ranchers sit down with environmentalists, business representatives, Native Americans, kayakers, hunting guides, county officials, federal and state land managers, and other concerned citizens. In addition, such efforts are iterative and ongoing as opposed to being single-play problem-solving efforts. Finally, GREM is initiated primarily, but not always entirely, by citizens and/or non-governmental entities like nonprofit groups (i.e., the impetus for action comes primarily from "below" rather than from the government "above").

GREM thus involves a dramatic shift in the organization and control of public bureaucracies responsible for managing the interaction between society and nature. Instead of centralized hierarchy, government experts in control, specialized agencies, and layer upon layer of written rules and procedures, GREM is premised on greater decentralization of governance, shared power among public and private actors, collaborative, ongoing, consensus-based decision processes, holistic missions (environment, economy, and community), results-oriented management, and broad civic participation.

Located principally in the Western United States, the new movement is rooted primarily in rural areas in which local economies are directly and inextricably tied to natural resources and now involves over 40,000 core participants and volunteers in over 500 communities.[8] Prominent examples include places such as Willapa Bay (Washington), the Malpai Borderlands (New Mexico, Arizona), the Henry's Fork watershed (Idaho), the Blackfoot River Valley (Montana), and the Applegate Valley (Oregon), although there is growing evidence that similar efforts are emerging in the Eastern and Southern United States, as well as across the globe (Born and Genskow 1999; Knopman, Susman, and Landy 1999; Lubell et al. 2002; Western and Wright 1994; Yaffee et al. 1996).

The attempts by citizens to reinvent governance regimes in the natural resources and public lands arena correspond with a broader movement in support of new paradigms for governance in a multiplicity of other policy arenas, including education, policing, rural development, public health, and tax administration.[9] All of this dovetails with what might be described as the national resurgence of an ethic of civic responsibility. In policy areas across the board publics are mobilizing against the perceived results of government regulation and, in some cases, the imposition of values contrary to their own. They are seeking to revitalize civil society, the intermediate realm of politics that lies between individuals and government, to reclaim the right to control, or at least to profoundly affect the substance and execution of public policies having the greatest impact on their lives and livelihoods.[10] In its most ideal form

civic innovation seeks to mobilize social capital in new ways, to generate new institutional forms, and to reinforce these through public policy designed for democracy. And it aims to provide citizens with robust roles—in their professional and nonprofessional roles, institutional and volunteer activities alike—for doing the everyday public work that sustains the democratic commonwealth. (Sirianni and Friedland 2001, 13)

As part of this new dynamic, many scholars and practitioners now contend that effective environmental policy programs require new rules of engagement that recognize the critical importance of social complexities and that strengthen the working relationships among diverse government, civic, and business actors at the state and local levels.[11] According to this view, the keys to enhancing environmental policy performance, and thereby improving the capacity of communities for achieving environmental sustainability, are greater, more substantive citizen participation,[12] collaborative decision processes,[13] and a new conceptualization of sustainability science that makes the case for sustainable communities.[14]

The Question of Accountability

At the heart of these debates over the performance of environmental policy and alternative institutions is the central question that drives this book—the question of democratic accountability (see Weber 1998, 232–235, 262–264). According to critics of the QLG, the rush to embrace collaboration, participation, and consensus imperils accountability to the broad public interest, particularly the national public interest and the interests of future generations. More specifically, critics argue that while the QLG's heart is in the right place, all it has really done is produce a logging bill disguised as a forest health management proposal. The QLG thus is an unwitting accomplice in the timber industry's strategy to gut national environmental laws, to limit public involvement, and to, more generally, establish a special interest governance "scheme . . . that relies on . . . excessive new logging . . . [and] provides scant extra environmental protection in return" (*San Francisco Chronicle* 1997; Cockburn 1997; Marston 1997b). It is "common among QLG opponents [to] refer to [it as a] well-intentioned Bambi consorting with [a] ravenous Godzilla, and naive chickens inviting sharp, high-powered foxes into the coop" (Mazza 1997, 3). The Sierra Club laments that the QLG solution is "an out-of-court, beyond-the-beltway solution to an intractable *national* issue" (MacManus 1997, 30). Felice Pace of Klamath Forest Alliance, a California-based environmental group, goes even further. She views the QLG as part of the larger conspiracy of globalization: "I cannot help but see this as the new international economy trying to co-opt the forest movement. Coercive harmony is a real phenomenon. . . . What better way [for Sierra Pacific, a timber company, to control policy than] to co-opt . . . local activists?" (as quoted in Mazza 1997, 3).

The critics of the QLG have quite a bit of company. Scholarly conventional wisdom is skeptical of the ability of decentralized, collaborative, participative arrangements to produce democratic accountability to broad public interests. One set of critics suggest that such arrangements may not be accountable given their propensity to produce agencies captured by private interests, co-opted national policy agendas, iron triangles, and, ultimately, policy outcomes benefiting the few at the expense of the many (Amy 1987; Bernstein 1955; Lowi 1979; McConnell 1966). Accountability occurs, but in zero-sum fashion—democratic accountability is limited to truly localized matters or to the preferences of private interests at the expense of broader state and national public interests. Past experiences in environmental policy provide considerable support for this perspective (Clarke and McCool 1996; Culhane 1981; Gottlieb and FitzSimmons 1991; Klyza 1996; Maass 1951; Selznick 1949). This, in fact, is the primary fear voiced by leaders of national environmental groups in their attacks on GREM. Michael McCloskey (Sierra Club) and Louis Blumberg (Wilderness Society) criticize the new arrangements as nothing more than an ingenious cover for the self-interested machinations of industry, who will use such proceedings to impose the values of economic growth and efficiency and to rid themselves of the burdens of national environmental laws (McCloskey 1996; Marston 1997c; see also Cortner and Moote 1999, 60; McGinnis 1999a, 500; Stahl 2001; Wondollek and Yaffee 2000, 230–233).[15] Others fear that the new community-based arrangements are simply a disguise for a more sophisticated Wise Use movement (audience reaction to presentation on GREM, Utah State University, November 1998).

A second stream of criticism is grounded firmly within traditional public administration theory as defined by the Progressive reform tradition and the classic Weberian legal-rational model of bureaucracy. GREM violates several of the cardinal precepts of administrative doctrine that are designed to ensure democratic accountability (e.g., includes citizen participation on a par with bureaucratic experts, breaches the sacrosanct public-private boundary, and dissolves hierarchical authority relationships in favor of shared power with nongovernmental stakeholders) (Moe 1994). There also are concerns regarding the difficulty of holding someone accountable and of ensuring performance in network-based arrangements. If all are in charge, then perhaps no one is in charge (Moe 1994)?

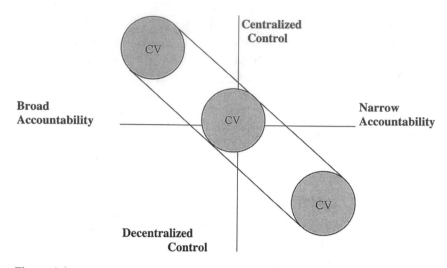

Figure 1.1
Accountability, the expectations of conventional wisdom

Likewise, "networks are [generally considered] weaker vehicles for social action" given the coordination problems stemming from the fact that all activity is jointly produced (Milward 1996, 79).

The conventional wisdom thus posits a largely linear relationship between centralized and decentralized control of governance arrangements and the type of accountability produced. The greater the degree of centralized control, hence less collaboration and direct citizen participation, the greater the probability that broad-based accountability (to nations and states/regions) will result, while decentralized control is most likely to equate with narrow accountability (to individuals and communities). Figure 1.1 depicts the expectations of the conventional wisdom.

Are the critics right? Is the QLG and, by association, other GREM efforts, dangerous to the public interest? Does a lack of accountability to broad public interests automatically accompany decentralized, collaborative arrangements that seek to bring society back in? The answers matter, not only from the perspective of American democracy, but also for the environmental policy realm. Because if the critics are right, if bringing society back in is synonymous with special interest government that has little regard for environmental protection, then most bets regarding improved program performance are off. In this scenario, GREM is likely to enhance policy performance for the select, powerful few, but at the cost of retarding progress toward environmentally sustainable communities that, by definition, promote a strong measure of accountability to future generations.

Conversely, what if the skeptics' conclusions are premature? Is it possible that the theory of accountability simply has been reconfigured to fit the new paradigm for governance in such a way that fealty to broad public interests is maintained? Put differently, rather than being dangerous, perhaps the conceptualization of accountability fostered by GREM is only different.

This book takes on the puzzle of accountability raised by these new governance arrangements by studying the actual practices of politics and policymaking, and documenting how actors come together in three cases of GREM—the Henry's Fork Watershed Council in east-central Idaho, the Applegate Partnership in southwestern Oregon, and the Willapa Alliance in southwestern Washington. It explores the characteristics and operational dynamics of the various accountability mechanisms being utilized in the new governance arrangements to find out how accountability works. Put differently, what are the ways in which accountability is operationalized when power has been decentralized and shared with the private sector, when the decision processes are premised on collaboration and consensus, when citizens actively comanage issues affecting public lands, and when broadly supported results are key to administrative success? And what are the conditions under which decentralized, collaborative, and participative policy administration arrangements are most likely to achieve accountability?

Another primary purpose of this research is to develop a better understanding of the connection between accountability and policy performance. This occurs along several dimensions in the book. First, unlike traditional conceptions of accountability, practitioners of GREM do not make a distinction between accountability and performance; policy results, or outcomes, are treated as an essential part of the accountability equation. Second, participants in these new governance arrangements expect to produce effective environmental policy that promotes positive-sum gains for the environment and economy. To address these issues, *Bringing Society Back In* examines thirty outcomes produced by the three cases of GREM to see if they accord with the claim of broad-based, simultaneous accountability and the goal of win-win outcomes for the environment and economy. Third, the environmental policy field is a special case as concerns accountability, and no analysis is complete without giving due consideration to a nonstandard, temporal dimension of accountability, namely, accountability to future generations, or the ideas of environmental sustainability and sustainable communities. Therefore, the book also explores the relationship between the practice of GREM and the goal of environmental sustainability.

In addition, the book is designed to fill a gap in several scholarly literatures by providing a comprehensive, detailed exploration of the accountability conundrum. To date, the public administration literature as well as the environmental policy and general public policy literatures have yet to address the question of accountability in the realm of alternative governance institutions except in limited fashion, or in a fashion ill-suited to the complex reality of the GREM phenomenon.

The book thus offers a rich description of what accountability looks like from the perspective of the participants, how it can be seen to work in these three exemplary cases of GREM, and the possibilities inherent in the practice of GREM with respect to the goal of sustainable communities. The book also provides a discussion of the conceptual linkages between accountability in theory and the practices of GREM. Just as importantly, it provides critical conceptual underpinnings for future empirical analysis, as found in the operationalization of accountability, the criteria by which policy and program outcomes are assessed, and the discussion of the conditions likely to promote accountability. At the same time, the cases are a way to test the conventional wisdom concerning the general lack of accountability associated with decentralized, collaborative, and participative institutional arrangements. Are these new governance arrangements dangerous or different?

The short answer, at least for these three cases of GREM, is that the accountability framework is different rather than dangerous. Contrary to conventional wisdom, the array and logic of the various accountability mechanisms employed in these cases suggest that these efforts can be simultaneously accountable to a broad cross-section of society—individuals, communities, surrounding regions, and the nation. (See figure 1.2.) Moreover, the thirty outcomes not only reinforce this conclusion regard-

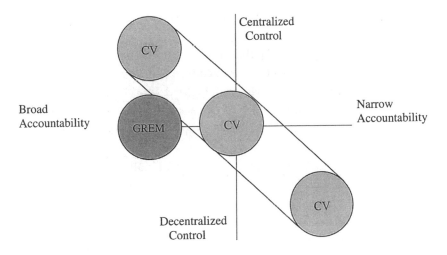

Figure 1.2
GREM accountability in relation to conventional wisdom (CV)

ing accountability, but because the outcomes are broadly accountable to a cross section of governmental jurisdictions, interests, and policy goals, including environmental protection and economic matters, they also lend support to proponents' claims that these collaborative arrangements can and do produce effective environmental policy.[16] Finally, the evidence associated with the three governance arrangements also suggests that the institutions, practices, and tools promoted by GREM can heighten and focus a community's sense of collective purpose on environmental sustainability, while concurrently adding the kinds of institutional capacities that help ensure progress toward the sustainable-community ideal.

What Is Accountability?

Accountability is a system, or set of mechanisms, designed to make sure promises are kept, duties are performed, and compliance is forthcoming. It implies that the person or entity being held accountable has an obligation or responsibility to an authority, group, standard, mandate, or behavioral norm *external* to the individual or organizational entity, or both. Hence, accountability implies control of behavior and the existence of an authority relationship between those being held to account and the entity or entities making sure accountability exists. Ideally, discretion is acceptable only if it does not interfere with accountability.

In the political science and public administration literature, democratic accountability is generally characterized in response to two questions— to whom? and for what? (Bardach and Lesser 1996). "To whom" accepts that accountability standards are often set or significantly influenced by political actors such as the president, Congress, congressional committee members, organized interests, states, or citizens. It recognizes that public policy programs are ultimately the product of political demands from various interests within the larger democratic system. The question of "for what" simply recognizes that a system of accountability is meaningless unless a substantive standard exists—the original promise or duty contained in an agreement, policy, mandate, and so on—against which to measure subordinates' performance.

The expectation is that if accountability is not forthcoming, political principals will exercise their authority to encourage more accountable behavior on the part of implementing agents. Examples include new written rules to clarify or modify expectations, or sanctions designed to motivate proper behavior (i.e., the successful performance of an obligation as defined by the principal). Widely known, general examples of political, or public, accountability relationships in the United States include the social contract between voter-citizens and elected officials, the formal constitutionally defined relationship between members of Congress, the president, and the bureaucracy, and the relationship between administrative leaders and subordinates inside public bureaucracies. In the first case, citizens have the right and the mechanism (voting) to remove politicians from office if they are deemed unaccountable or unresponsive to citizens' preferences. In the second case, the constitution vests significant control of the bureaucracy in Congress and the presidency. As long as they do not interfere with the Constitution itself, for example, legislators have the right and power to hold bureaucratic officials and their decisions accountable to the preferences of Congress using hearings, legislative authority, the power of the purse, and so on. In the third case, bureaucratic employees in the lower reaches of an agency may suffer transfers, demotions, or, in rare cases, dismissal for failure to properly discharge responsibilities.

Accountability as an Open Question

The definition of accountability makes clear that as a concept it is easily understood and its role in the successful practice of democratic governance is inviolable—end of story. Yet the definition does not dictate *how* to achieve accountability. Rather, it leaves open the possibility that a variety of different mechanisms are capable of solving the accountability puzzle for a given situation. Thus, while the concerns raised by the critics about accountability systems grounded in a devolved, collaborative, and participative style of governance *are* legitimate, it is not at all clear that their dire conclusions regarding a lack of accountability will obtain in the *contemporary* world of policy administration.

Recognizing this, an important contingent of scholars ranging across a broad array of literatures submit that the puzzle of accountability for decentralized, collaborative, and participative institutions is far more complicated than either conventional wisdom or contemporary critics make it out to be. They argue that we do not know the answer to the accountability question as it concerns reinvented government, especially decentralized, collaborative, participative arrangements. These scholars suggest that positive-sum or simultaneous accountability outcomes are possible (i.e., improved accountability to local interests, private and public, without a diminution of accountability to broader public interests), and that new frameworks are needed for understanding what accountability looks like.

Within the field of public administration, for example, Milward (1996, 89) and Kettl (1996) recognize that while collaborative policy implementation networks may make accountability far harder to secure by exacerbating problems of control, "the answer is not clear cut [concerning] . . . the accountability of [such] networks." Thompson and Riccucci (1998, 250–251) argue that given the potential for improved accountability within the reinventing government paradigm, it is now up to scholars to flesh it out. Bardach and Lesser (1996) take a few steps toward this end by clarifying the connection between the accountability "to whom" and "for what" questions, while suggesting, among other things, that the creation of behavioral norms for participants in collaboratives may help solve the accountability puzzle. Gilmour and Jensen (1998), Radin and

Romzek (1996), and Romzek (1996, 111) also accept that accountability is possible in the new world of policy administration. They each make the case that as the paradigm for governance and administration changes from top-down hierarchies to networks, devolved authority, and/or privatization, so too must systems of accountability. Behn (2001, 254) takes the discussion of accountability a step further by advocating a 360° accountability system focused on performance and potentially grounded in "a compact of mutual, collective responsibility . . . [where] every individual would, indeed, be accountable to everyone else." I say *potentially* because Behn only raises the theoretical possibility of mutual, collective responsibility, yet does not offer any empirical evidence where such a system is in operation. What is clear is that "traditional definitions of accountability are too narrow and restrictive to be useful in this dynamic [collaborative] environment" (Kearns 1996, xviii).[17]

Other literatures also suggest that accountability is possible with GREM. The *public bureaucracy* approach argues that differentials in the character, culture, and professional staffing of government agencies matter to policy outcomes, that bureaucracy is fundamentally political, and that the political contexts within which bureaucratic decisions are taken vary agency by agency (Katzmann 1980; Khademian 1992, 1996; Knott and Miller 1987; Wilson 1989). The variance in the internal characteristics of agencies and their external political contexts suggests that universal applications of administrative solutions will not elicit consistent outcomes. In other words, a traditional hierarchical approach to public administration may not provide accountability (e.g., see Light 1995; Behn 2001), while decentralization and collaboration do not automatically equate with a lack of accountability (see Behn 2001).

The emergent *social capital* literature also leaves open the possibility of accountability. To the extent that a community develops a web of horizontal, cooperative relationships built on trust, the more likely it is to demonstrate a capacity for effective self-governance, and the more likely it is to develop decentralized governance arrangements that are expressly designed to be accountable to a broad cross section of interests (Putnam 1993; Jackman and Miller 1998). Steven Rathgeb Smith's interest in nonprofit networking with public agencies brings him to the question of social capital as well. His work in *Public Policy for Democracy* (1993) emphasizes the importance of community capacity (broadly defined) for policy success, and hence the importance for public managers to have the skills for working with a variety of organizations and individuals outside of the governing organization. Robert Reich (1990b) and Marc Landy (1993), although not social capital scholars, come to much the same conclusion. To the extent that citizens participate in a public, civic process of discovering the public interest, self-governance is promoted and direct accountability is exercised.

Further, there is the *governance-without-government* literature, which discusses institutional arrangements governing the use of common property resources in the international arena as well as in small-scale, stateless societies.[18] According to these scholars, it is now well established that groups of interdependent actors, including those in the private and non-profit sectors, as well as within the public sector, can and often do succeed in managing the function of governance without resorting to the creation of governments in the conventional sense (Ellickson 1991; Milward and Provan 1999, 3; North 1990; Young 1996, 247; Young 1997, 5). Put differently, politically significant social institutions such as GREM can be (and are) effective at "resolv[ing] social conflicts, promot[ing] sustained cooperation in mixed-motive relationships, and, more generally, alleviat[ing] collective-action problems in a world of interdependent actors" (Young 1997, 4). Such arrangements are also capable of enhancing monitoring and enforcement capabilities (Ostrom and Schlager 1997). The broad message is that formal government institutions are not always necessary for achieving policy effectiveness and accountability.[19]

In a similar vein, the broader literature on *informal institutions* argues that accountability can be produced by informal institutions such as social norms. Norms create a specific set of behavioral expectations for members of a community, employees in an organization, practitioners of a profession, or participants in a particular institutional arrangement (Etzioni 1996, 1998; Fukuyama 1995; Katzmann 1980; Khademian 1992; March and Olsen 1989; North 1990). As Gormley (1995, 54) tells us, "if [norms are] well-crafted and well-diffused, [they] can substitute for formal structural controls." Bardach and Lesser (1996) suggest, among other things, that the creation of behavioral norms for participants in collaboratives may help solve the accountability puzzle. The social pressure for individual performance to match public commitments in a collaborative improves the quality of individual performance because

"the individuals kn[ow] they [will] have an audience for their performance, an audience that [can] be appreciative but that [can] also be critical" (205).[20] Braithwaite (1998, 351) goes so far as to argue that the
enculturation of trust within institutions is an informal, yet effective accountability mechanism because it helps "control abuse of power."

In each case, enforcing the behavioral expectations associated with informal norms is a matter of collective responsibility. Participants aware
of potential noncompliance generally rely on social persuasion, reminding the (potential) defector of their collective obligations, and warning
them of the possible consequences of their action (e.g., expulsion from
the institutional arrangement; loss of community "status"). In other
cases, violation of unwritten, unspoken norms may lead to the violator's
removal, often without explanation, from the informal, behind-the-
scenes decision-making loop, thus hampering the violator's ability to influence outcomes. Or, as in the case of policing, a senior officer may refuse
to offer standard assistance to their junior partner in times of trouble if
the "rookie" resists going along with established, yet informal depart-
mental "street" norms. The expectation is that such social sanctions are
severe enough that, over time, and in combination with the individual
"gains" from conformance (e.g., acceptance, inclusion, protection, influence), most people will embrace the norms of their group, organization,
or institution (Ostrom 1990).

In addition, there is *normative democratic theory* and the potential for
transformation through participation at the individual and community
levels. The work of Michael Piore (1995), John Dryzek (1987), Daniel
Kemmis (1990), Douglas Torgerson (1999), and others (Arendt 1959;
Barber 1984; Mansbridge 1980, 1990; Tocqueville [1835] 1956) suggests
that individuals regularly engaged in community deliberation or deliberative communication processes no longer see their preferences and priorities in strictly individual terms, but in the context of broader community
norms or structures. Who they are as an individual member of a community, hence their preferences for policy, unfolds in the context of their
participation with others in governance. Transformations are likely to
occur on two levels. Participation in these governing efforts might help
individuals to better see their relationship to others, including differences
and similarities of ideas (Warren 1992), or to identify primarily with part
of a larger group (Dawes, van de Kragt, and Orbell 1990). They might

begin to understand themselves in the context of community, in other words, rather than as autonomous individuals. Moreover, when individuals begin to see their own preferences in a broader community context, there are likely to be positive consequences for the overall capacity of a community to address *collective* problems (a performance benefit).

Moreover, given the emphasis on bringing society and its citizens back in to public policy decision-making frameworks, the character and competence of citizens play a more prominent role in the accountability equation. More specifically, *communitarians* argue that to the extent that a governance arrangement brings society back in, certain skills and character traits (virtues) are demanded of those citizen-participants if accountability is to be forthcoming (Glendon 1995). From this perspective, the challenge of public policy is "not just the manipulation of incentives but also the formation of character" (Galston 1995, 38). James Q. Wilson (1985, 16) agrees, "In almost every area of important public concern, we are seeking to induce persons to act virtuously. . . . In the long run, the public interest depends on private virtue." The communitarian literature goes on to suggest that (1) not all citizens possess the necessary bundle of skills and character traits required to promote a healthy and accountable system of democratic governance, and (2) some institutional arrangements are better than others at inculcating the kinds of skills (e.g., deliberation, compromise, consensus building) and virtues (e.g., civility, honesty, law-abidingness) demanded by such citizen-based democratic arrangements (Glendon and Blankenhorn 1995).

Finally, there is contemporary scholarship within the environmental policy literature suggesting that accountability to broad public interests might be possible for decentralized, collaborative, participative governance arrangements. Wondollek and Yaffee (2000) provide a brief discussion of how accountability might be accomplished using generally recognized principles such as performance standards, stronger monitoring and evaluation systems, and adherence to "fair" processes, while also concluding that "[their] experience examining close to 200 collaborative processes over the past decade suggests that the [critics'] fears [regarding a lack of accountability] have been realized on only a handful of occasions" (231, 237–242). In an earlier piece of research (Weber 1999a), I describe the systemic properties of GREM accountability as compared with four other identifiable political-administrative systems of accountability over the last 175 years of American history. I find that "the conceptualization of democratic accountability . . . varies dramatically over time. . . . Each conceptualization emphasizes different institutions and locates ultimate authority for accountability in differing combinations and types of sectors (public, private, intermediary), processes, decision rules, knowledge, and values" (453). Cortner and Moote (1999, 60), in *The Politics of Ecosystem Management,* barely scratch the surface of the accountability question and primarily revisit traditional concerns over the potential lack of accountability. Yet they also hold open the possibility that the "revis[ion] of social beliefs, values, norms, and governance practices" may well "resolv[e] the . . . paradoxes of decision making [that] occur between the goals of . . . inclusiveness and accountability," among other things (70). Pelkey et al. (1999), while not examining accountability directly, find in their study of 180 stakeholder-based, natural resource management efforts that the collaborative partnerships are *not* being formed "in response to demands from wealthy or ideologically motivated people," either liberal or conservative (3).

The shared theoretical thrust of these literatures is that both formal and informal (roles, norms, and customs) institutions matter, as does the larger political, economic, and social context within which institutions are set and within which individuals interact (e.g., Keohane and Ostrom 1995; North 1990; March and Olsen 1989). The expectation is that as institutions and their context change, constraints on behavior, opportunities for action (benefits), and the ground rules for social and political interaction are redefined. The redefinition of constraints, opportunities, and ground rules for interaction change behavior, either by altering the incentives affecting individual and group behavior (e.g., Knott and Miller 1987), or by transforming individuals' worldviews such that they start to think "beyond self-interest" (e.g., Mansbridge 1990), or both (a mixed-motive scenario). Of equal importance, the literatures suggest that the changes in behavior ultimately produce changes in outcomes.

If the theoretical approaches outlined above are right, the possibility arises that as the institutions and context of policy administration change over time, so too does the capacity for decentralized, collaborative arrangements to effect outcomes in keeping with a broader, rather than narrower, public interest. As part of this, there is also the possibility that the new governance arrangements will emphasize norms, rules, and practices promoting and strengthening accountability mechanisms, while simultaneously contributing to improved performance. These points are important precisely because *existing* natural resource management institutions are roundly and regularly criticized for being largely unaccountable, whether in terms of a broad public interest, financial mismanagement, the ability to produce on-the-ground results, or otherwise (e.g., Cushman 1999; General Accounting Office, 1998; Nelson 1995). Moreover, historical analyses offer particularly intense criticism of a specific kind of institutional arrangement governing natural resources—those predicated on the devolution of authority, *albeit in fragmented fashion*—given their propensity to produce policy outcomes serving the few at the expense of the many (Clarke and McCool 1996; Coggins 1998; Culhane 1981; Gottlieb and FitzSimmons 1991; Hays 1959; Klyza 1996; Lowi 1979; Maass 1951; McConnell 1966).

The problem is that even though accountability is possible in the new world of policy administration, we do not know what effective accountability looks like because there is a gap in the literature. While the importance of the accountability question is not in doubt—it *is* central to the study of public administration and public policy—the scholarly problem is that, with few exceptions,[21] the question of accountability, and the potential for the dilution of accountability, generally has been lost in the enthusiasm for reinvented government (Behn 1999; Durant 1998; Kettl 1996, 10; Romzek 1996; Thompson and Riccucci 1998, 254). As James Q. Wilson (1994, 668) notes, "the near absence of any reference to democratic accountability is perhaps the most striking feature of the Gore [reinventing government] report."

Scholars instead have devoted considerable intellectual attention to the (non)achievement of better government performance, and have typically examined accountability only within the narrow context of reinvented government personnel systems[22] and the legal ramifications of privatization initiatives (Gilmour and Jensen 1998). Or, as in a widely cited piece by Ronald Moe (1994), the new arrangements have been subjected to a scathing critique using a conceptualization of accountability that, by definition, finds them unaccountable. Others such as Khademian (1996), Light (1993, 1995), and Terry Moe (1989) focus on the question of accountability, but treat it in the traditional manner as a one-way street emanating from the top down (elected officials) rather than as a process of incorporating both top-down and bottom-up inputs (citizens) into the policymaking mix. Similarly, the vast majority of normative theorists seem content to confine their intellectual efforts to the realm of theory. They do not seek to operationalize and test their normative constructs in the contemporary world (although a few limited, yet notable attempts to bridge the gap between theory and practice have been made—for example, Warren 1992; Morrell 1999).

At the same time, scholarly treatments of the accountability question, even when focused on alternative institutions, tend to reduce accountability as a concept to analytically separate, often unlinked, and, in some cases, diametrically opposed pieces of the larger governance puzzle. Accountability becomes a matter of making sure, for example, that it is clearly focused on either the "top" or the "bottom," as opposed to any marriage of the two directions. Or the hallmarks for successful accountability within a particular system of governance become heavily weighted toward *single* factors like professional norms (Friedrich 1940; Radin and Romzek 1996), or proper institutional structure (Finer 1941; Wilson [1887] 1997), or appropriate written procedures (Gulick [1937] 1978), or citizens possessing enough virtue (e.g., communitarians). Within the reinventing-government genre, the focus has been more on performance-based accountability and all the problems associated with it—as if performance is the *only* way to measure and understand such a system of accountability.[23] Framed in this way, the discussion of accountability becomes mired in a discussion of which *individual* parts of the accountability solution are better than the others with respect to a particular governance paradigm, with no sense of the complex linkages between them, or the potentially fruitful combinations of the various component parts. Such a fragmented, mechanical approach is likely to be of limited value in helping us understand GREM-based accountability given its philosophical and practical embrace of a more complex, holistic, and organic approach to governance.

Regardless of whether the scholarly literature has yet to address the question of accountability in the realm of alternative governance institutions, or has done so but in a fashion ill-suited to the complex reality of the GREM phenomenon, the accountability conundrum associated with GREM is real. As the rhetoric surrounding the QLG's alternative proposal for governing and managing public lands makes clear, it is relatively easy to conjure up all kinds of horrifying scenarios of unaccountable or even special interest government under such circumstances. Given the more general trends toward the devolution of government power, the growing attempts by "reinventing" government enthusiasts to bring society back in, and the embrace of GREM by hundreds of communities, it is readily apparent that the horrifying scenarios of unaccountability *are* possible. Thus theory is needed to help improve our understanding of just what an effective system of accountability might look like so that such risks are minimized. As Thompson and Riccucci (1998, 254) aptly remind us, it is precisely because the

reinvention movement in American governance cannot appropriately be dismissed as folly, fad, or all talk and no action [and because] many of the propositions embedded in reinvention ideology are plausible working hypotheses about how to improve administrative performance and accountability, . . . [that] reinvention efforts should command the serious attention of political scientists as important instrumental and symbolic initiatives that may improve governance. . . . Taking reinvention seriously requires that scholars build a better knowledge base about the nature and consequences of administrative reform.

It is this lack of systematic empirical analysis of the accountability question and, by extension, the relationship between accountability and policy performance, that this research is designed to rectify.

Research Design and Case Selection

This book studies the actual practices of politics and policymaking in three cases of grassroots ecosystem management in order to explore whether accountability is possible for decentralized, collaborative, and participative institutional arrangements, and, if it is, to document what it looks like and how it works. The research is also designed to offer a preliminary discussion of the conditions under which these new governance arrangements are most likely to achieve accountability. In addition, the book investigates the connection between accountability and policy performance, or the actual outcomes being produced by these cases. Of specific concern are how the outcomes accord with any claims for accountability and the expectations of participants for environmental policy outcomes that are not only simultaneously supportive of the environment and economy, but also sensitive to the goal of environmental sustainability.

I settled on the question of accountability and picked the environmental policy field for a variety of reasons. In an earlier book on collaboration and environmental regulation in national pollution control politics (Weber 1998), I devoted considerable space in the final chapter to a discussion of accountability and collaboration, particularly the potential for the *lack* of accountability to a broad-based public interest (232–235, 262–263). However, I also noted that given the win-win character of the outcomes associated with the collaborative "pluralism by the rules" dynamic, the time was ripe for more empirical studies designed to help us ascertain more fully the validity of conventional claims concerning collaborative decision making and the lack of accountability.

At the same time that I was finishing *Pluralism by the Rules,* I became an avid reader of the *High Country News,* one of the American West's foremost sources of news on natural resource issues. The stories in the *High Country News* made it clear that the collaborative dynamic, whether in the form of watershed groups or community-based conservation, was picking up steam in communities across the Western United States because participants expected to produce outcomes supporting positive-sum gains for the environment and economy, while also contributing to participants' desire for more environmentally sustainable communities. Intrigued, I researched the new phenomenon and found that it was a new environmental movement—grassroots ecosystem management—with its own distinctive definition of environmentalism, preference for institutions and management approach, and approach to science and technology (see Weber 2000a).

The research into GREM also made it clear that the growth of these collaborative efforts in the environmental policy arena were setting off alarms concerning accountability within the national environmental advocacy community. National environmentalists expressed concerns regarding the potential for capture of these new efforts by extractive industry interests that, according to their analysis, would necessarily result in the acceleration of environmental degradation and an end run around national environmental protection laws. In addition, the rush to embrace new paradigms for governance, or reinvented government, across a broad swath of policy areas, including education, policing, rural development, public health, and tax administration, was raising similar questions related to accountability and policy performance.

The controversy over accountability, or the lack thereof, led me to explore the historical-comparative nature of accountability in American politics. I researched and reported on the broad outlines of GREM's accountability system as compared to four other major accountability systems over the past 200 years (Weber 1999c). The analysis found that acceptable systems of democratic accountability have taken a variety of forms through the years rather than adhering to some sacrosanct overarching notion of accountability, especially with respect to how to achieve it. The five different models of accountability—Jacksonian, Progressives/New Deal, public interest–egalitarian, neoconservative efficiency, and GREM—emphasize distinctive arrays of institutions and locate authority for accountability in differing combinations and types of sectors (public, private, intermediary), processes, decision rules, knowledge, and values.

The realization that the mechanisms of accountability can and do vary over time thus converged with the emergence of new governance institutions in a variety of policy fields and the contemporary public controversy over accountability to suggest an empirically based exploratory study designed to elicit new knowledge on the question of accountability. Given the extensive research that I had already conducted on GREM and the fact that these new governance arrangements are broadly representative of reinvention efforts in other policy areas, it made sense to focus on the environmental policy field.

At the same time, a sense of urgency underlies this research that is tied to the recognition that the world of environmental policy administration has changed in important ways (e.g., Kemmis 2001; Knopman, Susman, and Landy 1999; Mazmanian and Kraft 1999a; O'Leary et al. 1999; Wondollek and Yaffee 2000). The actual choices of collaborative governance arrangements being made by policymakers, bureaucrats, organized interests, and citizens are exerting a strong decentralizing effect, increasing the importance of organizational and program adaptability to changing and/or varied conditions, emphasizing interdependence as a prerequisite to public policy success, and adding complexity to public management arrangements by forcing people to come together and work across traditional interest and jurisdictional boundaries. When it is considered that the pressures on government agencies to cope with decentralization, adaptability, interdependence, complexity, and citizen demands for full participatory status are only likely to increase over the next several decades, the expectation is that successful governance will more and more become associated with an assortment of alternative institutional arrangements that defy traditional notions of regulatory and bureaucratic organization and control.

Moreover, we may not like, or we may rightfully fear, the burgeoning use of new governance arrangements that potentially threaten cherished ideals of liberal democracy, increase the risk of agency capture and special interest government, or place citizen input on a seeming par with policy mandates from elected officials. But, whether we like it or not, policymakers, administrators, and citizens *are not* waiting on scholars to decide whether such arrangements are appropriate. Seen from this perspective, because practice *is* running ahead of theory, the challenge to theory is real, not imagined, and we need to figure out what works and what does not. To the extent that scholarship ignores such developments, political science is hampered in its ability as a discipline to explain and assess, much less understand significant, interesting political-institutional phenomena such as collaborative decision-making arrangements.

Yet we lack a theoretical framework for understanding accountability within the context of collaborative efforts. In part this is because the conventional wisdom concerning decentralized, collaborative, and participative institutional arrangements argues that accountability is either not possible, or is at best highly unlikely (e.g., Culhane 1981; Lowi 1979; McConnell 1966). Yet there are also gaps in several scholarly literatures. To date, the public administration literature as well as the environmental policy and general public policy literatures have yet to address the question of accountability in the realm of alternative governance institutions except in limited fashion, or in a fashion ill-suited to the complex reality of the GREM phenomenon. Thus we are stranded, with little advice on how to proceed and little guidance on the positive aspects of alternative institutional arrangements. This book takes a first step in the direction of establishing a theoretical framework for understanding accountability in the case of decentralized, collaborative, and participative governance arrangements by taking James Q. Wilson's (1989, xii, 12) advice to dig into the empirical specifics of the phenomena in question.

Specifically, the research focuses on the practices and choices of exemplary, or successful, cases of GREM in order to find out what accountability looks like.[24] Each of the three cases—the Applegate Partnership, the Willapa Alliance, and the Henry's Fork Watershed Council—had been touted as clear cases of institutional success (with respect to decentralized, collaborative, participative arrangements) by the media, by numerous elected officials (both state and federal), and by affected administrative authorities.[25] The decision to focus on exemplary cases was made for several reasons.

First, there were data-collection concerns. If I picked a case of clear failure, it was not clear that participants would have been willing to talk about their experience(s). Second, it was reasonable to expect that if accountability was possible, it would be most likely to exist and to be recognized in cases of broadly recognized success. Such cases would offer a rich array of descriptive data across the full range of variables with which to paint an initial, yet reasonably accurate picture of an accountability framework or frameworks that could then be tested using a broader range of successful, mixed-success, and failed-attempt cases. Third, this research broadens the range of cases examined relative the existing literature by focusing on cases where success (i.e., accountability) is likely. The expectation is that the empirical record will cast light on what factors might contribute to the achievement of accountability.

Thus, the book describes what accountability looks like from the perspective of the participants, and how it works in these three cases of GREM. The book also develops the conceptual linkages between accountability in theory and the practices and choices of GREM participants. In addition, it provides critical conceptual underpinnings for future empirical analysis. And while *Bringing Society Back In* does not seek to be as rigorous as future research will need to be, it does attempt to provide the foundations for greater rigor in that research. The approach follows the advice of King, Keohane, and Verba (1994) and accepts that rich, descriptive "case studies . . . are . . . fundamental to social science. *It is pointless to seek to explain what we have not described with a reasonable degree of precision*" (44; emphasis added). In short, while the data provided by the exemplary cases do not offer definitive answers to the main research questions, they are a necessary first step for future research designed to build the kind of explanatory theory that will help us to know with greater certainty whether the Bambi-versus-Godzilla scenario proffered by critics or broad-based, simultaneous accountability is dominant across cases.

Further, for the purposes of an exploratory study of a descriptive question it is acceptable, methodologically speaking, to select on the dependent variable and to explore the dynamics of cases that are widely regarded as exemplary instances of a process or institutional arrangement (as in best-practice designs in organizational behavior research). The key, according to King, Keohane, and Verba (1994, 129) and Yin (1994), is that the complexity of these alternative institutional arrangements is such that there is a reasonable expectation of variation in how similar levels of success (the dependent variable) are accomplished across cases. In fact, this is precisely what the interview data show. The text and tables of chapter 3 illustrate the variation in what forms accountability takes (although there is considerable similarity, no case holds to one exact common form). Similarly, the actual outcomes being produced by the three cases of GREM vary in how accountable they are to the different levels of accountability—individuals, communities, states/regions, and nation (see chapters 4 through 6).

The research is important not only from the perspective of American democracy, but also for the environmental policy realm. If the critics are right, if bringing society back in is synonymous with special interest government that has little regard for environmental protection, most bets regarding improved program performance are off. In this scenario, GREM is likely to enhance policy performance for the select, powerful few, but at the cost of retarding progress toward environmentally sustainable communities. But if the critics are wrong, if broad-based accountability and improved policy performance are possible with the new governance arrangements, we are one step closer to offering much-needed systematic guidance to policymakers and public managers in the environmental policy field as well as decision makers in a whole host of different policy areas who are struggling with the same reinvention-accountability nexus. The payoff from their perspective is a better understanding of when they can devolve authority, bring society back in, and reap the benefits of reinvented government, while still preserving accountability to the broad public interest. We are also one step further along in the battle to clarify and illuminate the connection between complex, collaborative partnerships among diverse government, civic, and business actors, environmental policy performance, and the idea of sustainable communities.

Illuminating these relationships is also likely to help us better understand how to make and implement policy effectively in a world where decentralization, interdependence, and collaboration are defining elements. In advanced industrial democracies where there is limited political space for governance *without* government to take hold, the key may be hybrid institutional arrangements that combine governance *with* government in ways that strengthen or complement the existing system of government.

Finally, the study of accountability within the realm of environmental policymaking offers lessons for normative democratic theory by developing information on the potential for the transformation of individuals regularly engaged in deliberative communication processes. Do such people still see their preferences and priorities in strictly individual terms, as autonomous individuals separate from society, or does the participative experience lead them to understand themselves as more fully connected, and sensitive to, broader community norms or preferences? As with the other areas of inquiry, this research is only one step on the road to an answer; there is still a need for more systematic tests of this proposition in future research endeavors.

Research Methods

The research itself draws on original interview data, government documents, primary documents produced by the three GREM cases, and published accounts of the cases. A total of seventy interviews were conducted. Sixty-five of the interviewees were active participants in GREM proceedings. Twenty of these participants had attended GREM meetings for more than five years. Another thirty-two had been participants for more than two years but less than five, while the remaining thirteen participants had between one and two years experience with GREM meetings. Four interviewees were politicians or legislative staffers at the state and federal level with considerable familiarity with the efforts. One interviewee was a local politician who also was an active participant in one of the efforts.

Names were selected from records listing board members (the Willapa Alliance and Applegate Partnership cases), staff members (Willapa Alliance), and active participants (Henry's Fork Watershed Council; Applegate Partnership).[26] For example, the Applegate Partnership's board of directors is designed to reflect the diversity of the major stakeholding groups in the watershed. The eighteen seats on the board are allocated among nine major categories, with each category represented by two co-board members. The groups represented on the board are agriculture, environmentalists, unaffiliated citizens, timber-extraction (large) companies, small-tree farmers, local colleges, mining and mineral issues, manufacturing industries, and a local research organization, the Rogue Institute for Ecology and Economy (RIEE). The Willapa Alliance's seventeen-member board represents a diverse array of community interests and typically includes officials from large corporate timber companies, an Ecotrust representative, small-tree farmers, oyster and cranberry farmers, environmentalists, fishing interests, the Shoalwater Bay Tribe, and small businesses with stakes in recreation and tourism. In the Applegate and Willapa cases, at least one board member from each of the above-listed categories was interviewed. More generally, for all three cases participants were divided into twelve categories—environmentalists, extractive/commodity/business interests (e.g., mining, timber, agriculture, fishing, and so on), recreation, federal-level administrators, state-level administrators, local-level administrators, legislative/elected officials from all three levels of government, concerned citizens (no formal group affiliation), scientists (independent and university based), and media. At least one representative from each of these "general" categories was interviewed for each case, with the exception being elected officials (only five across the three cases) and concerned citizens with respect to the Willapa Alliance case.

In keeping with federal regulations regarding human subjects research and the requirements associated with prior approval for human subjects research at Washington State University, each interviewee was guaranteed anonymity. Further, given that the interviewees occupied positions that involve repeated interaction with others in government as well as within their own communities, many expressed concern about how their participation in a scholarly study of the GREM effort might affect ongoing relationships with other parties. Moreover, in some quarters (especially some government agencies) the use of collaborative, participative venues remains controversial. Consequently, the assurance of anonymity allowed numerous interviewees to discuss the various elements of the accountability framework with considerable freedom.

The primary interviews were conducted between March 1998 and October 1999. The interview sessions were semistructured but purposefully left open ended, requiring anywhere from forty minutes to two hours to complete. Follow-up telephone interviews for the purposes of clarification and fact checking continued through October 2000. Interview questions were designed to elicit information concerning two basic questions: Is accountability possible for decentralized, collaborative, and participative institutional arrangements? If so, what does accountability look like in a specific case? Participants were asked to define accountability, to provide examples to illustrate how accountability is being operationalized, and to explain why selected accountability mechanisms were both better and worse than existing arrangements. Interviewees were also challenged with critical statements suggesting that their administrative arrangements were, in fact, undemocratic and unaccountable. Moreover, interviewees were asked to explain why they were choosing to participate in the new institutional arrangements, as well as whether, and how, their participation was transformative. The actual interview format is displayed in appendix A. The interview data collected were critical to the development of the accountability framework set forth in chapter 3, the detailed discussion of thirty outcomes contained in chapters 4 through 6, and the background materials found in chapter 2.

Reconfiguring Accountability to Fit the New Paradigm for Governance

How have the Henry's Fork Watershed Council, the Applegate Partnership, and the Willapa Alliance reconfigured the theory of accountability to fit the new paradigm for governance? And where is the evidence that suggests that these new governance arrangements produce a form of accountability that merits the label of broad-based, simultaneous accountability (i.e., accountability to a broad cross section of society, the nation, and the region as well as the community)?

The solution offered by the three cases of GREM to the accountability conundrum is not simply structural or performance based or procedural; instead it is multifaceted, complex, and holistic in its approach. It is about communities—defined as "communit[ies] of place . . . not simply a summation of the residents of a particular area, but rather . . . a set of social relations that develops in relation to a place"[27]—making a credible commitment to a conceptualization of democratic accountability that is broad (collective) and simultaneous. At the most basic level, participants are embracing a series of social, political, and administrative institutions, "or sets of roles, rules, [norms,] decision making procedures, and programs that serve to define social practices and to guide interactions of those participating in these practices,"[28] which, in theory, promote such a form of accountability. The five key components of the broad-based, simultaneous accountability framework are as follows (table 1.1 lists elements for each item):

• Formal institutional structure
• Formal institutional processes (procedures)
• Informal institutions—participant norms, the enculturation of virtue, and a credible commitment to broad-based accountability by leaders
• An ecosystem (holistic) approach to management
• A focus on results (performance)

The new governance arrangements thus not only rely on accountability mechanisms grounded in more traditional formal institutions such as institutional structure and institutional processes (procedures), they employ a distinctive array of additional *informal* institutional mechanisms in pursuit of broad-based accountability. In addition, participants in GREM accept that a key part of the accountability equation centers on government performance, or the ability of government to actually deliver on promises. The twin goals of improved governance performance and democratic accountability are usually treated as involving zero-sum trade-offs; enhancing one diminishes the other. Yet GREM tries to enhance both at the same time. As such, the idea of governance performance is not separated from democratic accountability, but instead becomes an essential component of the overarching conceptualization of accountability.[29] Participants argue that without an explicit focus on performance and on-the-ground policy results, accountability fails to capture the essence of the democratic process. Performance adds meaning to the deliberation and conclusions emanating from political institutions by explicitly connecting promises with outcomes.[30]

The complex array of, and synergistic interaction between, the five main accountability mechanisms evident in the structure and operation of GREM appear to promote broad-based (collective), simultaneous

Table 1.1
Elements of the accountability framework

		Sources of accountability		
Formal institutional structure	Formal institutional processes	Informal institutions	Ecosystem (holistic) management	Focus on results (outcomes)
• Collaborative, nonhierarchical (horizontal) design • Direct, participatory decision making • Role and position relative to others —Catalytic, coordinative core —Advisory character • Place-based character • Open-access design • Open, public information systems • Legal charter	• Joint deliberation and negotiation, and repeated interaction • Consensus decision rule • Standard procedure for major decisions • Community-building exercises • Oversight/monitoring processes • Enforcement process	• Participant norms (e.g., inclusiveness, civility, honesty, etc.) • Enculturation of virtue • Credible commitment to broad-based accountability by leaders	• Holistic, integrated approach • Broad knowledge base • Flexibility, adaptability, experiments • Proactive problem solving	• Participants insist it is essential to larger accountability equation • Needed because no "one best way" to achieve accountability • A critical tool for assessing whether the promise is matched by reality

accountability in a number of ways. First, the institutional arrangements pose significant obstacles to political interests seeking to impose their *individual* will on other GREM participants or to craft outcomes favoring a narrow set of interests or values.

Second, GREM management arrangements inculcate allegiance to the value of broad-based accountability. For example, the design of the new governance arrangements suggests, and field interviews confirm, a normative concern for outcomes responsive to a broad range of people and institutions. Put differently, participants are concerned with "who benefits?" Without this emphasis, the possibility exists that both responsiveness and performance will be forthcoming, yet the end result will be policies designed to benefit the few at the expense of the many. As part of this emphasis on decisions reflecting a broad public interest, GREM specifically incorporates sustainability—the consideration for future generations—into its mission and decision-making dynamic. It thus embraces a specific style of results orientation, a holistic approach to the notion of environmental protection and management with the expectation that doing so will promote sustainable ecosystems, communities, and economies.

Third, the new arrangements recast problems, opportunities, constraints, and relationships by changing incentives for action. The changed incentives for action translate into changes in behavior and outcomes (both of which are choices).

Finally, in at least some cases, the dynamic associated with GREM appears to transform individual citizens and how they view their role in the broader system of governance. Transformed citizens are more likely to (1) view issues from the perspective of enlightened self-interest or even from the broader perspective of the community and/or region as a whole, (2) embrace the idea that citizenship carries with it certain responsibilities, including the obligation to equip oneself with the skills necessary for the practice of self-governance, and (3) accept the value of virtuous behavior as a viable, necessary part of the remedy to the accountability equation.

Saying that certain sources, or mechanisms, of accountability exist and explaining how each element works (the primary task of chapter 3), however, says nothing about the degree of assurance that broad-based accountability will actually occur. The final component of the new accountability framework—a focus on results (outcomes)—thus comes into play, in part because participants claim it as an essential part of the larger accountability equation. Yet the focus on outcomes is also important because it takes us beyond a superficial concern with written rules and organizational structure as proxies for the real thing (i.e., accountability), while simultaneously recognizing that in a world of devolved, shared authority designed to incorporate the bottom-up concerns of a multitude of actors, there is unlikely to be one best way to conceptualize or construct a viable, democratic administrative system of accountability. Accepting accountability in such a robust form, however, increases the uncertainty surrounding the actual achievement of accountability, particularly from the perspective of elected officials *outside* the new governance system (e.g., Scholz's "political game" level, 1991). Because the new governance arrangement is nested within the larger political game played at state and federal levels of government in the United States, the uncertainty must be resolved to politicians' satisfaction.[31] Otherwise, it is unlikely that those with political authority will grant the discretion required for the new institution to function effectively over the long term.

To the extent, then, that the analysis of results demonstrates a positive relationship between the accountability framework and the decisions produced by GREM, it will help elected officials, as well as public managers within affected agencies, determine the strength of an accountability system and whether and when devolution should occur. In this way, the results component of the accountability framework provides a critical tool for assessing whether the promise of broad-based accountability suggested by the institutional dynamic is matched by the reality of the outcomes being produced. Without it, even if the larger model is coherent and logical, the analysis runs the risk of being nothing more than an intellectual fantasy—good for the ivory tower, but having little to do with political, social, environmental, and economic reality, and thus of limited value to citizens, policymakers, and bureaucratic practitioners seeking to improve governance.

The evidence associated with the outcomes suggests that, once again, GREM can and does produce decisions consonant with a conception of accountability that is broad-based and simultaneous. This is not to say that decentralized, collaborative, and participative governance arrangements will always produce broad-based accountability, but that contrary to conventional wisdom, such an outcome is possible with the proper mix of institutions, leadership, and management approach. Seen in this light, perhaps the best way to characterize GREM is as an ongoing experiment trying to come to grips with what Altshuler and Parent (1997, 11) call one of "the great contemporary challenges of public management, . . . [the ability] to demonstrate effectiveness rather than mere adherence to rules, a capacity to learn rather than just mastering established routines, and democratic accountability by means other than action 'by the book.' "

Prior to the full articulation of the new framework of accountability and the examination of outcomes, chapter 2 develops the context surrounding the emergence of the new governance institutions in three areas of the rural West—the Applegate Partnership in southwestern Oregon, the Henry's Fork Watershed Council in east-central Idaho, and the Willapa Alliance in southwestern Washington. In addition, the backgrounds and landscapes of the communities and places in question are described, and the actors in each drama are identified.

Chapters 3 through 7 rely on the Henry's Fork, Applegate, and Willapa cases of GREM. The Henry's Fork Watershed Council is a state-chartered watershed council in Idaho involving land that is 50 percent public and 50 percent private. Two groups—irrigation (farmers) and environmentalists—that had been long-time adversaries joined with federal and state land and water management agencies, ranchers, local officials, and other citizens to create the cooperative council in 1993. With broad support from Idaho's congressional delegation, Clinton administration officials such as Bruce Babbitt (Interior) and Mike Dombeck (U.S. Forest Service), and Idaho's legislative and executive branches, the Council has been widely touted as a promising prototype for reinventing and improving natural resource management, while simultaneously reengaging citizens. Former U.S. Senator Dirk Kempthorne (R-ID; now governor) and U.S. Senator Mike Crapo (R-ID) have even promoted the Council at the national level as a new, proactive way to prevent species from becoming endangered (see Kemmis 2001).

The Applegate Partnership, located in southern Oregon's Applegate Valley, was initiated in 1992 by a broad array of concerned private citizens in tandem with federal public lands agencies. In part because the BLM and the USFS own approximately 70 percent of the land covered by the Partnership, Secretary of the Interior Bruce Babbitt visited the Partnership in 1993 to gather ideas for the upcoming Timber Summit in Portland, Oregon. The success of the visit convinced him to hail the Partnership as a model of future forest management at the Summit, to ask two of the Partnership's board members to participate in the Summit, and to model several of the Summit's Adaptive Management Area recommendations after the Applegate initiative (Moseley 1999; Shipley 1996, 3; personal interviews, 1999).

The Willapa Alliance is a GREM effort in southwestern Washington, just north of the mouth of the Columbia River. Coinitiated by nonprofits and private citizens in 1992, the Alliance's success in coordinating and catalyzing a comprehensive fisheries management plan for the area has prompted state legislative leaders, as well as Washington Governor Gary Locke, to propose the Alliance's administrative format as *the* template for rescuing threatened natural resources, like endangered salmon runs, throughout the rest of the state. Former Washington Governor Booth Gardner even has gone so far as to label the Alliance "an exceptional process of trying to create good jobs that can be maintained for generations to come . . . [and that] may serve as an example to the world" (*Chinook Observer* 1996; personal interviews, 1998). The Alliance has also received national exposure through a laudatory National Public Radio report in 1997 and citation as an exemplary case of a sustainable communities initiative by President Clinton's Interagency Ecosystem Management Task Force (1996).

Chapter 3 provides a rich description of what accountability looks like from the perspective of the participants, and how it can be seen to work in these three cases of GREM. Put differently, the chapter fleshes out the institutional specifics of the broad-based, simultaneous-accountability framework, while also explaining the logic behind each mechanism and denoting the differences among the three cases. Chapters 4 through 6 then explore a total of thirty outcomes produced by the three GREM governance arrangements to see if they accord with the claim of broad-based, simultaneous accountability and the goal of win-win outcomes for the environment and economy. Chapter 4 is devoted to the Applegate Partnership case, chapter 5 covers the Henry's Fork Watershed Council, and chapter 6 investigates the Willapa Alliance. The investigation of outcomes for each case is organized according to four criteria—the diversity of representation in processes and individual outcomes; the relationship of choices (outcomes) to existing laws, regulations, and agency programs; the extent to which choices benefit a broad array of interests; and the specific effect of outcomes on "individuals" within the community. A second level of analysis maps the same outcomes against four different levels or scales at which accountability might occur—the individual (micro), community (meso), state/region (mid-macro), and nation (macro). If the common criticisms are valid, there should be a large imbalance in the way accountability is apportioned among the levels. Most outcomes should end up strengthening accountability to the individual and/or community (meso) levels at the expense of the state/region and national levels. On the other hand, outcomes that meet the demands of a broad-based, simultaneous form of accountability will exhibit accountability to multiple interests and levels at the same time. The empirical evidence associated with the thirty outcomes supports the general claim of broad-based, simultaneous accountability, and, because the outcomes are broadly accountable to a cross section of governmental jurisdictions, interests, and policy goals, including environmental protection and economic matters, they also lend support to proponents' claims that these collaborative arrangements can and do produce win-win outcomes for the environment and economy.

Chapter 7 explores another, less traditional level of accountability that is nonetheless of central concern in environmental policy circles. This is the idea of environmental sustainability, a perspective on accountability that factors in the effects of institutions and decisions on future generations of citizens. A key reason participants are buying into these new institutional arrangements is the expectation that their efforts will be rewarded with sustainable ecosystems, communities, and economies, especially when compared to the outcomes produced by existing institutional arrangements. The problem is that we currently possess extremely limited information regarding how these new governance arrangements, much less others, actually translate the goal of sustainability into reality.[32] Using the same three case studies, I find that GREM efforts can promote the kinds of institutions, practices, and tools that heighten and focus a community's sense of collective purpose on environmental sustainability, while concurrently adding the kinds of institutional capacities that help ensure progress toward the sustainable-community ideal.

The final chapter, chapter 8, develops the implications of the findings for scholars and practitioners in public administration, political science, natural resources management, and environmental policy, more generally. Of chief importance is a discussion of the conditions under which broad-based, simultaneous accountability is most likely. In other words, how can we best guarantee accountability to the diverse interests involved and to the different levels of government? The insights derived from studying accountability are also used to inform the debate over sustainable communities by suggesting hypotheses related to its successful practice. In addition, the chapter uses common criticisms leveled against decentralized, collaborative, and participative institutions to examine potential weaknesses of the new framework for accountability, but does so while simultaneously turning a critical eye toward the weaknesses of existing accountability mechanisms. Further, there is the question of whether the new governance arrangements are more of a supplement or complement to existing institutions than a complete replacement for them. Does GREM fit the mold of a governance-*without*-government model, or a hybrid "governance-*with*-government" model? The chapter subsequently explores the main areas where GREM strengthens or complements the existing system of government. Finally, the chapter poses a series of questions suggesting next steps in the research agenda, both in terms of the question of accountability and the nascent efforts to create environmentally sustainable communities.

2

Rural Communities Embrace Grassroots Ecosystem Management

For those living in crowded, congested cities and suburbs where population densities commonly reach 2,000 to 3,000 people per square mile or more, it is probably hard to imagine the vast, largely unpopulated landscapes of the rural American West. Yet rural Western areas rarely exceed a density of twenty to twenty-five people per square mile (Riebsame and Robb 1997, 55). For those living and working in walled urban canyons and concrete jungles, it is doubtless hard to imagine the endless horizons, majestic vistas, sheer rock massifs, and towering forests of green common to much of the western United States. For those familiar with the West of storied legends—of conflict, of hardship, of wildness and vigilante justice—and of contemporary timber, water, mining, and fish (dams and salmon) wars, it is probably even harder to imagine communities of disparate interests sitting down and resolving their differences without guns, threats, arrests, or violence of any kind.

Yet in a growing number of areas across the Western United States, communities endowed with few people—but tremendous stores of natural resources—are sitting down together and cooperating for the sake of preserving and, in some cases, restoring their special place. The Henry's Fork Watershed in Idaho, the Applegate Valley in Oregon, and the Willapa Basin in Washington represent three such cases of cooperative grassroots ecosystem management. What is perhaps most remarkable about the Henry's Fork and Applegate communities is that it was only yesterday—in the 1970s and 1980s—that conflict raged, tires were slashed, and gunshots rang out. The Willapa Basin, on the other hand, watched the natural resource wars spread across the West and concluded that it was probably in for much of the same unless preemptive actions were taken.

This chapter develops the context surrounding the emergence of the new governance institutions in the Applegate, Henry's Fork, and Willapa areas. It also seeks to familiarize the reader with the landscapes and backgrounds of the communities and places in question, to outline the basic common form and missions of each effort, and to identify the actors in each drama. In each case, communities that had been torn apart, or feared being torn apart, recognized and reacted to the costs of the status quo by pushing for the creation of innovative new governance arrangements predicated on decentralization, collaboration, and bringing society back in.

The Henry's Fork Watershed

The Henry's Fork of the Snake River is about as far away from the Pacific Ocean as any drainage in the larger Columbia River Basin gets. Located in eastern Idaho and nestled up against Yellowstone and Grand Teton National Parks, the 1.7-million-acre Henry's Fork watershed, with 3,000 miles of streams and irrigation canals, boasts mild summers and difficult winters, in which it is not uncommon to see temperatures dip below zero by 30° to 40°. The "signature" of the Henry's Fork area, however, is its view: "The [eastern] horizon is interrupted by the glistening massif of the Grand Teton, rising from the high plan and stabbing the heavens like an unsheathed stiletto. It is a disorienting sight, looming over this landscape of well-tilled farms and meandering creeks" (Durning 1996, 275–276). The watershed encompasses three counties[1]—Fremont, Madison, and Teton—and is now home to 40,000 people (Henry's Fork Watershed Council, 1998, 1).

The first Euro-American settlers arrived in the upper Henry's Fork watershed during the 1860s and devoted themselves to hunting, trapping, cattle ranching, and commercial trout fishing on Henry's Lake. By the late 1870s, Mormon farmers heeding Brigham Young's call to make the arid region bloom, began settling the lower watershed, despite annual rainfall measuring a scant 20 inches. The Mormon settlers quickly tapped the irrigation potential of the Henry's Fork, building a network of canals to support the primary crop of seed potatoes. By the turn of the twentieth century, bountiful fish stocks at Henry's Lake, primarily native trout (cutthroat), combined with the impressive scenic vistas and an abundance of wildlife to spawn further settlement by wealthy industrialists from the eastern United States and a thriving commercial fishery rivaled only by the salmon fishery at the mouth of the Columbia River.

Over the next sixty years significant segments of the watershed came under the control of federal and state agencies such that land ownership is now split evenly by public and private entities.[2] During the same period, numerous state and federally sponsored water storage, recreation, and hydroelectric projects were constructed, including the dam at Island Park, Idaho, and the ill-fated Teton Dam, which in 1976 washed away under a torrent of rain, causing the loss of fourteen lives. Timber production in the Targhee National Forest, established in 1908, became an economic mainstay of the area in the decades after World War II, while the highland forests and meadows continue to support abundant deer and trophy elk as well as smaller populations of moose, wolves, and grizzly bears. Meanwhile, fish hatcheries, both private and public (operated by the Idaho Department of Fish and Game), introduced tens of millions of nonnative rainbow trout into the watershed, so that rainbows now dominate the main stem of the Henry's Fork as well as most of its tributaries. The success of the fish-stocking effort gradually earned the Henry's Fork watershed a reputation as one of the premier rainbow trout fisheries in the entire world (Van Kirk and Griffin 1997, 253–259).

The Applegate Valley and Watershed

The Applegate Valley and watershed, often simply called "the Applegate," encompasses almost 500,000 acres in southwestern Oregon (and a tiny part of California). The river drainage feeds into the Rogue River, which flows quickly out through the Coastal Mountain Range and into the Pacific Ocean. Approximately 12,000 people live in the Applegate, with almost everyone living in the rural, unincorporated areas of Josephine and Jackson counties.

The Applegate watershed is steep and rugged, with elevations ranging from 1,000 to 7,000 feet. Northern slopes, especially in the highlands, tend to be heavily forested with many remaining stands of old-growth and late successional reserve forests, while southern slopes typically produce brushy, small-diameter woody growth like Madrone and Manzanita trees. The lowlands and riverbeds offer prime soils for farming and grass for grazing (cattle ranching).

As part of the Klamath Geological Province formed 250 million years ago, the Applegate experiences great genetic diversity (biodiversity) because the geological formation provides a "bridge" for plant migration between the Cascade and Coastal mountain ranges. The bridge is still functioning today for scores of rare plants (e.g., Baker's cypress and American saw-wort), sensitive vertebrates (Townsend's western big-eared bat, bald eagles, and peregrine falcons), and unusual parent rock types (e.g., serpentine). The Applegate also has the largest number of known spotted owl nests of the ten federal Adaptive Management Areas created by President Clinton's Northwest Forest Plan in 1994 and is home to prime salmon (coho and chinook) and steelhead habitat. Moreover, there are more than fifty different soil types and a variety of microclimates where rainfall varies from 20 to 100 inches per year. In short, "there is no other comparable area on the Pacific coast with as much biological diversity" (Applegate Partnership, 1996a, 8; Sturtevant and Lange 1995, 7–8).

Beginning in the late 1800s, timber harvesting, farming, ranching, and recreation became the mainstays of the local economy, with many current residents tracing their lineage all the way back to nineteenth-century settlers. Early in the twentieth century, with the designation of most of the Applegate as federal public land, most of the timber and recreation activities came under the control of the U.S. Forest Service and the U.S. Bureau of Land Management. In fact, 70 percent of the Applegate's 500,000 acres are federal. The USFS manages 39 percent of the total acreage under the aegis of the Rogue River National Forest, while the BLM manages 31 percent of the land through the Ashland and Grants Pass Resource Areas (both are in BLM's Medford, Oregon District) (Rolle 1997a, 612–613).

In the 1960s and 1970s, and then again in the 1990s, two new waves of inmigration occurred, each consisting of "urban escapees," and each challenging the established cultural and economic patterns of the area. The first wave involved counterculture types and "back to the landers" interested in escaping a hedonistic, materialistic world by living off the land. The second wave involved "amenity" migrants, primarily from California. Amenity migrants are people who trade their wealth and, in many cases, their knowledge-based skills for a region's safety, slower pace of life, and relatively unspoiled natural landscapes.[3] Many of the amenity migrants are self-employed, connected to markets and business centers through mail-order catalogs, fax machines, and the Internet (Sturtevant and Lange 1995, 8).

The Willapa Basin

The Willapa goes by other names—shoal-water, ring of rivers, rain country, and a tidewater place, to name a few. All the names belay the essential characteristic of the Willapa—water. . . . Water is the ultimate currency in Willapa. . . . It is an essential element to every commercial activity. . . . Water is common property. It is the element that connects the seemingly disconnected activities in the Willapa. . . . A majority of the problems, challenges, and possibilities that arise in the Willapa are related to water and will increasingly become so. (Fred Dust, Willapa Educator)[4]

The Willapa Basin sits among the coastal hills of the Pacific Ocean just north of the mouth of the Columbia River in southwestern Washington State. The basin stretches over more than 1,000 square miles (about the size of Rhode Island) with over 100 miles of shoreline and an estuary, Willapa Bay (150 square miles), that is the West Coast's second largest, after San Francisco Bay. It "is the cleanest estuary of its size on the West Coast, perhaps in the continental United States. A huge tidewater surge floods the shallow bay every 12 hours. Some 25 million cubic feet of water are flushed into the Pacific Ocean at each ebbing tide. . . . [As a result,] every two weeks, the water in the bay is completely replaced" (Hollander 1995a, 2). The 80,000-acre bay is tucked into a landscape of steep interior canyons, marshy rivers, and myriad feeder streams in which the dominant habitat is rain forest, with Douglas fir and western hemlock as principal tree species. Other habitats include dune and sea-cliff grasslands, coastal pine forests, extensive salt and freshwater marshes, Sitka spruce swamps, and high-elevation grasslands called *balds*. Located on the Pacific Flyway, the Willapa Basin is also a major feeding and resting area for millions of migrating shorebirds and waterfowl (Willapa Alliance, 1998d, 1).

The region's economy is powered by a diverse array of industries such as forestry, fishing, tourism, and the farming of oysters, cranberries, and dairy and beef cattle—industries that share a common dependence on natural resources and on a clean environment (Allen 1992). The north end of the region around South Bend and Raymond, Washington, traditionally has been logging country, while Bay Center and Nahcotta, Washington, are the oystering hubs, and the flat river-bottom lands are home to ranchers and farmers (Hollander 1995a, 3).

Land tenure in the Willapa Basin is largely a private affair. Eighty percent of the Basin's 600,000 acres is privately owned, with federal holdings at a minimum (roughly 3 percent). Farmers and other small landowners own about 100,000 acres of the land, but it is the timber companies that own the bulk of the land. Weyerhaeuser Corporation, the international timber company, owns and manages 286,000 acres, or almost 48 percent of the landscape, while the Campbell Group, another timber company, is the second largest in the area with close to 100,000 acres in holdings (Hollander 1995c, 9; Manning 1997).

The $25-million-per-year oyster industry in Willapa Bay produces "nearly one-fifth of the nation's oysters" and its cultivated cranberry bogs provide about 10 percent of the domestic market demand for cranberries (Allen 1992). Its "succulent, . . . famous shellfish are ranked among the top in the nation," in part because it has maintained its water quality while "other great oyster producing bays—like Maryland's Chesapeake . . . —are paying the price of population growth" (Hollander 1995a, 2). A number of fish species also breed in the watershed, including three varieties of salmonids—coho, chinook, and chum or dog salmon—as well as steelhead and sea-run cutthroat trout. Today, however, most of the fish stocks, with the exception of chum salmon, are hatchery raised.

Though for most of the past 150 years the "green gold" of timber and healthy fisheries have dominated the economy, since the late 1980s the Willapa area has suffered significant job losses in the natural resource sectors. Today, the 20,000 year-round residents are experiencing a "a second era of discovery" led by tourists and retirees, who are the primary reason in-migration now exceeds out-migration. In fact, the influx of retirees has been such that one in five citizens are now over the age of sixty-five, compared to only one in ten in 1940 (Hollander 1995d). Moreover, tourism is now the Willapa's biggest industry, with seasonal jobs up 20 percent over the last ten years. Visitors fish, kayak, walk on the white-sand beaches, attend festivals (e.g., art, music, and kite flying) and, more generally, come for the solitude and rustic charm offered by the coastal, rural atmosphere (Hollander 1995f, 16).

Communities Defined by the Hatfields and McCoys

The health of the communities and economies of the Henry's Fork, Applegate, and Willapa Bay areas is clearly dependent on nature and its resources. Virtually everyone in each of the three places experiences a common, direct connection to the natural landscape; it is inescapable. As Kemmis (1990) explains, the basic ingredients required for the practices of cooperation and community based on the politics of place are in abundance. Yet the reality of governance has been something else indeed, at least in recent decades.

Beginning in the 1970s and 1980s, debates over the management of natural resources in each of the watersheds took on a visceral, shrill tone with neighbor pitted against neighbor and compromise a commodity in rare supply. By the early 1990s, the conflict had escalated to the point that some wondered whether "the fabric of community [that had previously existed] had been torn completely asunder, never to be replaced" (personal interviews, 1998 and 1999). In the Henry's Fork watershed, there were tire slashings, death threats, and arm-in-arm marches by thousands of farmers and ranchers down Main Street in St. Anthony, Idaho, to intimidate environmentalists and the growing legions of flyfishing enthusiasts who sought changes in the water-flow regime of the Henry's Fork for the sake of improved fish habitat. People thought to be sympathizing with the "other" side were pulled out of church in the middle of services and publicly rebuked "in order to straighten them out" (personal interviews, 1998). One author sums up: "Environmentalists painted irrigators as welfare cowboys slopping at the public trough. Resource developers painted conservationists as citified misanthropes bent on socialism" (Durning 1996, 276).

In the Applegate, while the USFS and BLM clear-cut huge swaths of forests during the 1980s trying to meet the probable sale quantity (PSQ) targets set by agency managers, a number of local environmental advocacy groups, spearheaded by Headwaters, pursued litigation and direct action with equal zeal. Tree spiking, a willingness to lie down in front of logging trucks (human roadblocks), and face-to-face shouting matches with local loggers were not uncommon. The success of environmentalists' tactics was such that by the mid-1980s most timber sales on public lands in the Applegate and throughout the region had been halted. In addition, the 1990 federal court decision granting endangered status to the northern spotted owl halted most public land management activity in the Applegate (Sturtevant and Lange 1995, 8; also see Durbin 1996). A local environmentalist from the Applegate Valley explains:

Polarization is a word I hear a lot when people try to explain what was happening to timber communities along the Pacific Coast. But it's too mild [a term] for the Applegate. You have to realize that for most of the 1980s, we were a community at war. . . . It was not uncommon to end up at the Country Market down in Murphy, [Oregon] standing in line with somebody that works for Boise Cascade [a timber company], somebody that works for the Forest Service, and an environmentalist. And all of a sudden an argument would erupt and before you knew it, they would be beating each other silly. . . . It was so divisive that you . . . were either with us or against us depending on what side you were on. It was like the Hatfields and McCoys. (personal interview, 1999)

Another longtime resident of the Applegate, a farmer, adds: "Logging-truck drivers . . . carr[ied] guns to work because they were being shot at. . . . It was really scary . . . , we were afraid somebody was going to get killed" (personal interview, 1999).

By contrast, citizens involved in the founding of the Willapa Alliance acted out of concern for the "gathering storm clouds of controversy" and "to preemptively head off the environmental wars . . . happening across the Pacific Northwest. . . . The sky [had not] fallen on [the Willapa Basin] . . . at least not yet . . . not completely" (testimony of Dan'l Markham, Executive Director, Willapa Alliance; U.S. Senate, 1997, 50–51). Rather than worrying about actual listings of endangered species, residents were primarily concerned about "the great challenges to our communities" posed by the potential Endangered Species Act (ESA) listings of spotted owls, marbled murrelets, and salmon (Markham, U.S. Senate, 1997, 48). The aim was to stop things from escalating to the point of fistfights and gunplay. Nonetheless, the intracommunity tension was still palpable and much greater than anticipated, especially in the early years of the Alliance. A local environmentalist notes that

there was a great suspicion, apprehensiveness, a fear that [the Willapa Alliance] was a sort of Trojan Horse that had some greenies [environmentalists] stuck inside of it, and that we were going to come out and take over the world by locking up and regulating lands and stopping people from using the land as they had in past generations for hunting, fishing, and other types of recreation. We also encountered opposition from fishermen and timber folk, which is the bulk of the economy, jobwise, because they feared, and I think legitimately so, a loss of income. (personal interview, 1998; Hollander 1995c, 8)

Recognizing the Costs of the Status Quo

By the late 1980s and early 1990s, it had become increasingly apparent to many of the citizens and stakeholders in each of the three communities that conflict and reliance on adversarial advocacy tactics (e.g., litigation, lobbying, mass rallies) were not only destroying any sense of community, they were not necessarily beneficial to long-term ecological and economic health either. In short, current institutions were increasingly viewed as doing a poor job of providing for environmental sustainability, economic stability, and community stability/sustainability (see also Lead Partnership Group, 1996, 36). More specifically, current institutional arrangements were criticized for being "poor problem solvers," for producing suboptimal outcomes or even for exacerbating environmental problems, and for alienating the public. Additional concerns centered on growing development pressures threatening environmental quality and established ways of life. Coupled with these realizations was growing recognition that virtually all community stakeholders depended directly on healthy ecosystems and that failure to take a broader stance regarding threats to the ecosystems in question would likely hasten degradation (Born and Genskow 1999). Taking the final step toward new governance arrangements, however, required a push from community leaders, and, in one case, from organizations outside the immediate geographic area. These individuals and groups finally recognized that "enough was enough," and decided that the time had come to explore the possibilities of a different way of doing business.

Working Alone, Just Saying No, and Unintended Consequences

Participants in GREM believe that too often, current politics rewards a self-interested orientation and, correspondingly, reinforces winner-take-all or win-lose solutions for problems affecting the community, especially those deriving from judicial decisions. An environmentalist opines that "it was either the environment was going to win or the economy, but not both together. It was way too simplistic. Owls versus jobs. Fish versus people. And the way we were approaching it as a society was mutually exclusive. I don't believe in that. . . . There's nothing that is mutually exclusive" (personal interview, 1999). Another environmentalist observes that "when one side wins the rest of the community is often left out in the cold" (personal interview, 1998), while others, including a small-business owner, a farmer, and a representative of recreation interests, argue that such an approach means that broad public interests are not addressed, at least from the perspective of the community (personal interviews, 1998). Nor does such an approach necessarily help with problem solving, according to Michael Jackson, the leading environmentalist from the QLG:

I can't fix the salmon problem with the law. They're in too much trouble. I need the help of everyone. It's the same with logging. I need my neighbors. If the logger who drives the Cat in the woods won't help me, then that tractor will go through the stream, no matter how many rules there are. (as quoted in Marston 1997c, 1)

At the same time, in areas with large tracts of private property, GREM participants believe that solving environment, economy, and community problems requires active, voluntary partnerships with current landowners. A leader from a state-level environmental group captures this sentiment; he maintains that "there is no way that you can buy up all the land to provide the requisite levels of environmental protection, you need to capitalize on local pride in place and build local capacity if sustainability is going to work" (personal interview, 1998).

Similarly, participants are tired of national interests whose modus operandi seems to revolve around "just saying no"—no more development, no more logging, no more dams, no more cattle, no more people. They point out that "success in stopping stuff is not the same as success at creating solutions. It is easy to say no, it's really difficult to get out and figure out what we are going to do to make something work better for all concerned"(personal interview, 1999).[5]

Moreover, even when single-issue interests adopt a more hands-on approach to problem solving in combination with traditional adversarial tactics, progress toward intended goals can be tricky. For example, by the early 1980s, a general consensus had developed in the angling and environmental community that rainbow trout numbers in the Henry's

Fork River below Island Park Dam (the Box Canyon and Harriman Ranch sections) were decreasing. Idaho Department of Fish and Game surveys corroborated the decline; in 1978 there were almost 19,000 rainbows in the Box Canyon, by 1983 there were only 15,000. In response to the decline, the Henry's Fork Foundation (HFF), an environmental organization, installed solar-powered electric fencing along the streambank from the mouth of the Box Canyon downstream 12 miles. The fencing prevented cattle from degrading the streambank, thereby minimizing the negative effects on fish habitat from excessive silt loads in the river. The HFF also successfully lobbied to institute more restrictive fish harvest regulations (catch-and-release fishing) and to protect the watershed from further hydroelectric and irrigation developments.

All three efforts improved or maintained existing fish habitat and increased the prospects for fish survival. Yet the rainbow trout population continued to decline. In 1989, one year after catch-and-release fishing rules were applied to the Henry's Fork, the rainbow trout population in the Box Canyon area totaled only 5,000 fish. By 1991 the Box Canyon population had fallen to 3,000 fish (Kirk and Griffin 1998, 263–264). Standard adversarial interest group tactics and fish recovery strategies were not working. As a member of the HFF explained, "it was clear that there needed to be something different because the same old way of doing business was not going to get us anywhere, was not going to help us solve the resource problem" (personal interview, 1998). Environmental advocates were winning the battles but losing the war because the ultimate results were the opposite of what they were striving for—a healthier environment and robust fish populations.

Another key concern is the unintended consequences stemming from inadequate information about current conditions—ecological, economic, social—and the lack of mechanisms for evaluating results (a measurement-and-information problem) (Willapa Alliance, 1998b; personal interviews, 1998 and 1999). A study by the Lead Partnership Group (1996, 35) of Northern California and Southern Oregon found that years of intensive management activities on public and private lands have produced "a variety of unintended consequences including loss of biological diversity, dramatically increased risks of fire and an overall decline in forest health. Similarly, community development activities, particularly those focused on rural communities, have taken place without clear as-sessment of current conditions, community-generated vision or system-atic evaluation of the effectiveness of intervention." The acquisition of more information—especially more accurate information—and the de-velopment of more effective monitoring mechanisms are viewed as meth-ods for empowering the community and aiding decision makers at all levels of government. Accurate and more robust data sets about current conditions and trends "will . . . enable [people] to make informed deci-sions about the parameters of community health and well-being." Simul-taneously, "the lack of effective assessment and monitoring programs perpetuates the opportunity for interest groups to create narrow defini-tions of community well-being or economic health" (Lead Partnership Group, 1996, 37–38).

Adversarialism as a Community Buster

The ongoing destruction of the community fabric produced by continual fighting is a key reason many citizens are joining the new governance efforts. They see the new institutional arrangements as a strategy not just for counteracting the "negative social residue" outlined above (i.e., death threats, gunplay, and so on), but also for building community anew, for restoring and maintaining civility, friendships, and the special character of their place (personal interviews, 1998 and 1999). An environmentalist and member of the Applegate Partnership explains: "I have spent a lot of time in Mexico and Central America and what I found there was com-munity and people who rely upon one another, particularly in rural com-munities. And so that is something that was lacking in our lives. There was no community. . . . And so part of what drove me to get involved was that I saw the [effort] as a way to try to redevelop the community we had lost" (personal interview, 1999). Fighting and litigating also can detract from problem solving. People end up being "pit[ted] . . . against one another instead of against our problems. . . . Our goal becomes de-feating the enemy rather than improving, restoring, and reviving our damaged ecosystems. And if either side ever does win, it will find its re-ward to be a hollow one, since in the very act of winning it has created a losing side that vows to become stronger and take its turn as the victor next time" (Daggett 1995, 8).

The Mismatch between Current Institutions and "Wicked" Problems

Since the Progressive era of the late nineteenth and twentieth centuries, the preferred form of bureaucracy has emphasized specialization in personnel (experts), tasks, and policy domains. Legislators and policy analysts typically treat the policy domains, for the purposes of management, as distinct, separate problem spheres in which policy-specific problems are identified and resolved. In addition, the assessment of progress is typically confined to the narrow specifics of the policy domain in question. Thus the USFS manages the nation's forests, while the BLM does likewise for public rangelands and mineral holdings (with some forests), the EPA wrestles with pollution control, and so on. Moreover, at the federal level and in many states, especially Western states adhering to the prior-appropriation doctrine,[6] water quality as a policy issue is treated separately from water quantity.

The fragmentation extends deep into such agencies and is generally defined or, at minimum, is reinforced by legislation. For example, the EPA treats problems as if they were discrete (media-specific) and has separate divisions devoted to air, water, and land (solid-waste) pollution. Success, according to the air office, occurs when *air* pollution is reduced or eliminated for a specific pollutant or in a specific class of sources even if the solution worsens water or land pollution problems. The USFS is guided by many pieces of federal legislation and a longstanding ethos among professional foresters that defines and interprets the mission very narrowly in favor of "getting the cut out" (timber production) with little or no regard for biodiversity or other ecological goals (Wilkinson 1992). In these ways, current institutions are known for producing the classic bureaucratic dysfunction called *trained incapacity,* a general inability to address problems from a comprehensive or "big-picture" perspective, and a problem-solving style that eschews cooperation and integration (Davies 1990; Knott and Miller 1987; Nelson 1995; Weber 1998).

Participants in GREM, however, focus on problems such as environmental sustainability, ecosystem health, and community stability that are systemic (i.e., deeply embedded in human nature and the social fabric), require ongoing attention (they are not amenable to a one-time solution), and cut across established legal and administrative jurisdictions (Mathews 1999, 81; personal interviews, 1998 and 1999; Community-Based

Collaboratives Workshop, 1999). Such "wicked" social, economic, and environmental problems are a poor match for existing institutions and are not susceptible to easy cures (Mathews 1999; Osborne and Gaebler 1993). Mathews (1999, 81–82) elaborates: "The usual strategy of breaking the difficulty into subcategories, designing categorical programs for each part, and holding one institution [or one set of experts within a particular institution] accountable for the 'solution' is as ill-suited for dealing with this kind of problem as putting a cast on someone suffering from diabetes would be. The remedy doesn't fit the disease. Wicked problems require action by the whole of a community."

Fostering a Loss of Trust in Government

Participants in GREM criticize government as inaccessible, biased, inefficient, and ineffective. The perception is that existing participation processes are not fair because they are dominated by organized interests and tend to place too much emphasis on science and expertise and not enough on social/community impacts and needs. Moreover, information is hidden or lacking, which creates information asymmetries favoring government experts at the expense of affected citizens (Levi 1998, 88). Henry's Fork watershed residents complain about the "staged," "rigged," "symbolic," and "secretive" character of traditional public participation processes such as public hearings and advisory councils (personal interviews, 1998 and 1999). A local farmer says that "people want to feel involved in the process [of governing]. You don't feel that with government processes [where] you can say what you think, but you don't know whether your comment is heard, or how decisions are made" (Johnson 1995a, 9). Nor is there a great deal of trust between the community and federal agencies in the Applegate. An environmentalist notes that "distrust in the BLM and the Forest Service is a huge factor. . . . Historically speaking, the public involvement component of federal agencies has lacked legitimacy. It is seen more as a process to confirm foregone [agency] conclusions" (personal interview, 1999). Federal officials also recognize the problem. An official who is a participant in the Applegate Partnership explains:

One of the biggest problems the Applegate Partnership must overcome is the lack of trust between the community and BLM. When we put our projects together we have to do massive sales jobs. . . . [For example,] on the Thompson Creek landscaping project . . . we were visiting people, calling people saying, "You've got to believe me, the BLM is trying to do something really different. You know I've lived in this area for over twenty years, the BLM has really screwed up in the past. But I'm telling you there's new kids on the block here and they're trying to do something different with this project. I'm sure that we're not going to see clear-cuts out here, and we can guarantee that we're going to see something that really enhances the watershed." Hearing that from a community member is a lot different than from a BLM person, because no one would believe them. (personal interview, 1999)

The lack of trust is so profound across the Pacific Northwest that it is not uncommon to hear the Bureau of Land Management called the Bureau of Land *Massacre,* and the U.S. Forest Service labeled the U.S. Forest *Dis*service, especially by environmentalists (Durbin 1996).

Over in the Willapa Basin, residents tend to focus their ire on state agencies, given the small federal holdings. One resident, a small-business owner, complains that "at one time [state agencies] were going to forget about this part of the world because they could not control it. You'd call them about something and you'd never get an answer. . . . And the way the laws were written. You could only have one [state] law, and they'd write in temperature standards and other water quality measures that didn't always fit the Willapa" (personal interview, 1998). A local elected official adds that "the state agencies are not in a position where they can gain the confidence of the community. They're outsiders living in Seattle and Olympia making decisions for the local [natural] resources and us. . . . They seldom do anything efficiently. They're really good at spending an awful lot of money without getting much accomplished on the ground" (personal interview, 1998).

Community Crises
Momentum toward a new institutional framework for managing natural resources and recreating common ground, or "some semblance of community," gathered steam when crises, potential and actual, helped to focus residents' attention on the problems associated with existing institutional arrangements.

The Henry's Fork watershed suffered two serious river-related crises in the summer and fall of 1992. In both cases, the sudden introduction of massive sediment loads created havoc within each riverine ecosystem, causing traumatic shock to existing fish populations and suffocating insect beds on river bottoms (which are critical sources of food for birds, fish, and other aquatic species). In the first case, Marysville Hydro Partners, Inc. received Federal Energy Regulatory Commission (FERC) approval to build a small hydroelectric plant on the Falls River and to enlarge an existing irrigation canal to carry water to the plant. The FERC license was granted and construction begun amidst controversy over the adequacy of public involvement in the decisions and the accuracy of the data produced by the environmental impact assessment. A construction accident in June 1992 then released 17,000 tons of sediment into the river below the dam (Kirk and Griffin 1998, 264–265). The second crisis, or "disaster," occurred in September 1992 when the federal Bureau of Reclamation and the Idaho Department of Fish and Game introduced toxins (rotenone) into Island Park Reservoir in order to kill "trash" (non-game) fish, a common treatment for eventually restoring populations of more popular game fish like trout. Accomplishing the task required a drawdown of the reservoir's water level beyond normal end-of-season levels. Yet the agencies, focused as they were on the primary task of fish restoration, failed to understand or account fully for the ramifications of the drawdown on other watershed resources. Exposure of the reservoir's silt bed flushed massive quantities of fine sediments (50,000 tons) downstream into the Henry's Fork River over a two-week period. Agency officials were not even aware of the problem until an individual citizen with years of experience watching river conditions, both as a resident and as a fisherman, alerted them to what he perceived to be unusually high sediment loads (personal interview, 1998). Several eventual Henry's Fork Watershed Council participants from the environmental and recreation communities took the incidents as further lessons in how government agencies, rather than being accountable to all, "lack[ed] accountability to many of the groups in the watershed directly affected by the decision" (personal interviews, 1998).

Rather than experiencing dramatic single events, the Applegate watershed saw two issues—injunctions against logging and forest fires[7]—become primary catalysts for change during the late 1980s and early 1990s. Injunctions related to the spotted owl restricted logging and decreased logging volume on public lands in the Applegate area. Less activity in the woods translated into lower employment levels for mill workers and loggers, declining sales at local businesses, and declining federal-to-county tax receipt transfers from timber activities on BLM Oregon and California Railroad (O & C) grant lands and USFS forests. Historically, one-half of the receipts from timber sales on the O & C lands and fully one-quarter of government receipts related to timber sales on federal public domain lands have been returned to the counties from which the sales originate. Such funds provided the majority of financing for a myriad of county services to Applegate residents, including the sheriff's patrol, the District Attorney's Office, health clinics, juvenile protection and detention programs, road maintenance, public schools, and the Rural Action Team Station in Ruch, Oregon (U.S. Bureau of Land Management and U.S. Forest Service, 1998; Hannum-Buffington 2000, 78). The combined effect of the logging injunctions in the Applegate were severe enough to "cause both businesses and residents to identify the 'timber' issue as the number one issue facing their community" in the early 1990s (Hannum-Buffington 2000, 78; Reid, Young, and Russell 1996).

Forest fires were a double-edged sword in the Applegate. Citizens both feared fire and welcomed it as a potentially useful management tool for restoring ecosystem health. Within the Applegate watershed, fire has helped the ecosystem evolve for thousands of years. While major fires occurred naturally every seven to twenty years, frequent low-intensity, short-lived fires regulated forest density, creating open, parklike stands of large-diameter trees and promoting species diversity (U.S. Bureau of Land Management and U.S. Forest Service, 1998). As a result, the Applegate is classified as a "fire-dependent" ecosystem "with 28 different species of conifer, over 400 species of brush, and a thousand different grass species, all of which are fire dependent" (personal interview, 1999; U.S. Bureau of Land Management, 1999; Applegate Partnership meeting minutes, July 7, 1993). But federal fire-suppression practices throughout the twentieth century succeeded in keeping the Applegate fire free since 1920—"the longest fire-free period in the Applegate watershed in 300 years" (Hannum-Buffington 2000, 83).

The policy successes, however, caused changes in the forests that threatened the ecological integrity of Applegate-area forests. The changes "dismantled . . . many of the ecological processes that sustained their productivity and resiliency" by producing extremely dense riparian and upland vegetation patterns "two to five times greater than would be expected to maintain healthy stands of trees" (Hannum-Buffington 2000,

83; U.S. Bureau of Land Management, 1999; U.S. Bureau of Land Management and U.S. Forest Service, 1994). Citizens were among the first to notice the changes, according to a federal official:[8]

[Community] residents were saying what the hell is going on in our woods? You know we're concerned about all these Pine trees dying . . . and it looks like the big Doug[las] firs are dying too. . . . Is it just a fluke, or the beginning of the end? Will we look like Eastern Oregon and the Blue Mountains in another twenty years with just huge swaths of big dead trees? . . . And we said, sure enough, . . . not only are they dying, they're dying at an alarming rate. We're heading down a disastrous road if we don't start doing significant [tree] thinning and reintroduce fire. (personal interview, 1999).

At the same time, the dense, overstocked, high-fuel-load forests increased the risk of intensely hot, catastrophic fires (as opposed to the historical pattern of regenerative, low-intensity fires). Catastrophic fires were more likely not only to destroy the property of those living in the area, but to increase the brittleness of forest ecosystems by damaging soils, completely consuming the forests, diminishing critical foraging supplies for deer, and causing stand (tree) replacement, which diminishes biodiversity (Hannum-Buffington 2000, 84). A key participant in the Applegate partnership observes that "the millions of acres of fires across the West and Northwest in 1987 and 1988 were a major wake-up call for us. And I think that collectively we started to understand that the ecological damage [wrought by] these catastrophic fire events was not acceptable. I mean, where was the wisdom eighty years ago when we decided to start messing with the ecosystem by snuffing out fires? We put ourselves on kind of a crash course with ecological disaster. . . . We violated the pact with nature" (personal interview, 1999).

The Willapa Basin, on the other hand, was facing a series of economic and environmental problems. The entire area had long suffered from high rates of unemployment, with rates averaging approximately 4 percentage points higher than the overall Washington State unemployment rate from 1970 through 1993. In the early 1990s, unemployment rarely fell below 8 percent and regularly climbed into the double digits during low-season winter months. During the 1980s, the story was even worse. On an annual basis, unemployment averaged 13 percent (Hollander 1995g, 16; Willapa Alliance, 1996g, 15; Willapa Alliance, 1998d, 25). Personal income per capita mimicked the unemployment story as high-paying manufacturing jobs, many of which were associated with the timber industry, started giving way to lower-paying service-sector jobs in the tourism sector. While personal income rose from $10,000 to $14,000 on a per capita basis from 1981 to 1989,[9] Willapa-area wages lost ground vis-à-vis the rest of the state. In 1981 Willapa residents earned 86.2 percent of the state average. In 1989, they earned only 77.3 percent of the broader average (Willapa Alliance, 1996g, 18–19). Poverty measures also showed a community in growing distress. The share of people living under the poverty line in 1980 was 10.9 percent, only slightly higher than the state-wide average of 10.3 percent. Yet by 1990, the Willapa Basin poverty rate jumped to 17.2 percent, significantly higher than the comparative 10.3 percent state average (Willapa Alliance 1996g, 18–19).

In the environmental arena, as previously noted, there was widespread concern in the Willapa Basin about the effects of potential ESA listing of various wildlife species on the community and economy. Others perceived that the fragmented government bureaucracy was neglecting pressing local problems, such as the depletion of chum salmon runs and the rampant growth of invasive, nonnative vegetation (e.g., spartina) that negatively affects shellfish reproduction. In addition, citizens feared that eventual bureaucratic solutions (if any) would be zero sum in character, choosing to close down fishing grounds, shellfish-harvesting flats, and forests, with little attention given to the longer-term needs of a community that prefers a "working bay" and "working forests" (Hollander 1995g; personal interviews, 1998). Spencer Beebe of Ecotrust, a nonprofit organization in Portland, Oregon, articulates the fear: "The federal government lists the owl or some salmon run and sends out $50 million worth of welfare checks two years later. We are trying to dig in and find alternatives" (Blumenthal 1997). The alternatives, according to a local environmentalist, ideally would help residents maintain their dignity and, in as many cases as possible, their traditional livelihoods (personal interview, 1998).

Development Pressures

There is equal concern over the pressure to develop the Applegate, Willapa, and Henry's Fork regions for tourism and recreation purposes, and as newly discovered havens for trophy vacation and retirement homes. Such development pressures feed into the desire of citizens to build on the cultural and economic conditions of local communities, to create an indigenous capacity for adapting to changes in the world around them, and to maintain the ability to control their own future. To capture the trophy home dynamic, citizens of the Henry's Fork watershed have coined a saying: "The big millionaires (billionaires) in Jackson, Wyoming, are forcing the little millionaires over the Teton Mountains into the Henry's Fork watershed." As a result, property prices in prime areas around Island Park, Idaho (the upper Henry's Fork watershed), have tripled since 1990, while property valuations in the areas of Teton County closest to Jackson have increased at an even faster rate (personal interviews, 1998). Similar property-value increases have become common in the Applegate Valley as a new breed of farmers has ventured into wine production, and as newcomers have constructed homes that, as one federal official put it, "take a small forest to build" (personal interviews, 1999; Priester and Moseley 1997).

On the one hand, many citizens see these development pressures as a clear threat to ecosystem health and, by extension, to economic viability and the prized character of their place. More homes equal more solid waste and more septic/sewer problems negatively affecting water quality. For all three cases examined here, this is a problem for fish survivability as well as for human health problems (excessive levels of pollutants in well-based water supplies). For oyster farmers and cranberry growers in the Willapa, in particular, water pollution problems likely spell economic disaster given the inseparability between water quality and eventual product quality (Hollander 1995d; Allen 1992; personal interviews, 1998). More people also means more cars and RVs, more roads, greater air pollution, and "more of nature being gobbled up" (fewer open spaces). From this perspective, new development makes explicit the "limits" of existing natural resources while simultaneously highlighting the environmental value of maintaining existing land-use patterns. According to this view— a view supported by several environmentalists (participants) in the three cases of GREM studied here—the worst farmer or cattle rancher is still better than a new housing or minimall development (personal interviews, 1998 and 1999).

On the other hand, farmers, ranchers, and loggers as well as other long-time residents see development as a direct threat to their established "culture" and quality of life—greater traffic congestion, transient populations of distant property owners importing the frenetic pace of urban/suburban lifestyles and associated amenities (e.g., strip malls, fast-food restaurants, multiplex theaters), higher levels of transients associated with a tourism- and recreation-based economy, more crime, and changing values translated into changing politics placing greater restrictions on the uses of, and access to, land. They want to preserve traditional lifestyles and livelihoods directly connected to the land. Stan Clark, a representative for then–U.S. Senator Dirk Kempthorne (R-ID) who lives in the Henry's Fork watershed, expresses this sentiment well:

> As the sun came over the Tetons this morning, I was out on my horse. There isn't anything that teaches me who I am better than that horse between my legs, and good cows and calves in front of it, and looking at those things [the mountains]. . . . We have a way of life that should be preserved, and I think it can be preserved with the diversity of people [at this meeting of the Henry's Fork Watershed Council]. (as quoted in Durning 1996, 278)

Moreover, the dramatic rise in property values threatens to accelerate the rate of development and disrupt the local culture even further by displacing those longtime residents of modest means (with low and/or fixed incomes) unable to afford the higher property taxes. Others caught between the pincers of low commodity prices and the potential for enormous riches from the sale of their property may voluntarily leave the community by cashing in (personal interviews, 1998, and 1999).

Enough Is Enough: Leadership Helps Pave the Way for Cooperation and Common Ground

Leaders and concerned citizens in all three communities—the Henry's Fork, Applegate, and Willapa—finally decided enough was enough. In the Henry's Fork region, the two chief adversaries, Jan Brown, executive director of the HFF (representing environmental and flyfishing enthusiasts) and Dale Swenson, executive director of the Fremont-Madison Irrigation District (supplier of water to 1,700 farms in the region), "both had taken on larger-than-life images with their then adversaries" (Johnson 1995a, 9). Exhausted and frustrated from constant battling over the disposition of natural resources and land management approaches, Brown and Swenson decided to sit down and see if there was an alternative, more amicable method for reconciling their differences. They wanted to see about exploiting the common ground that they were sure existed (but nonetheless was routinely ignored given the historical animosity among participants, institutional fragmentation, and so on) for the purpose of improving governance effectiveness in the Henry's Fork watershed. The exploratory sessions led to the formation of an ongoing, cooperative, consensus-based decision forum involving the stakeholders and citizens of the watershed region.

Over in the Applegate, a similar dynamic was engulfing Jack Shipley, an executive board member of Headwaters, a local environmental group, and Jim Neal, a logger specializing in aerial or helicopter logging (and codirector of the Aerial Forest Management Foundation). With the spotted owl court decision literally shutting down timber extraction in area forests and with no seeming compromise in sight, or at least none that would simultaneously take into account the human needs of community members and the needs of the environment, Shipley and Neal called together a group of sixty community members in the fall of 1992. Meeting on the deck of Shipley's house, participants, including federal officials from both the BLM and USFS, hammered out a sketch of what a new institution for managing the lands and resources of the watershed might look like. They found that "there was considerable overlap between the desires and interests of environmental and industry groups, centered on maintaining the long-term health of the watershed and stability of local economies" (Shipley 1996, 1–2). Yet, while the vast majority of participants came because they agreed with the Shipley-Neal vision for a cooperative, watershed-wide effort, some came out of

fear. . . . Jack Shipley walked into my [agency's] office and unfolded a large map and said, "Look, here's the drainage . . . we're going to all manage this, this giant amoeba that he had outlined in yellow, together. . . . We won't worry so much about administrative boundaries, we really do what's needed for the land and the best thing for the people here." Well, [as a government agency,] we weren't even thinking in terms of watersheds yet. It was a new idea. And I looked at that map and it encompassed pretty much the whole area I was responsible for, well over 100,000 acres. It was like, "Oh my god, I've got to go watch out where the hell these folks are going." (personal interview, 1999)

In the Willapa Basin, however, the alliance received much of its initial impetus from external agents. In 1989, Dick Wilson, an oyster farmer and twenty-year resident of the Willapa area, met with Spencer Beebe, a scion from an old-money Portland family, over beers to discuss sustainable development in rural communities. Beebe's "work with ecologically sound development" in various parts of the world over the previous decade had left him wondering about and interested in the "implications of sharing the ideals of conservation with people deeply rooted in the landscape that surrounded them" (Hollander 1995c, 8; Hollander 1995e). Wilson, for his part, knew that there "was untapped potential in the people and places of the immense estuary and its rugged hills" that likely could contribute to preserving the pristine character of the Willapa's water for future generations (Hollander 1995c, 8).

Beebe went on to found Ecotrust, a nonprofit organization focused on sustainable development, in 1990. In 1992, Ecotrust then joined with the Nature Conservancy—a national conservation organization specializing in protecting the environment by purchasing and managing land of important ecological value—as well as with Dick Wilson, Dan'l Markham, and other Willapa Basin residents, to found the Willapa Alliance. The alliance was designed as a collaborative with a sustainable-communities mission that brings all of the bay's industries, communities, and interest groups to one table. Initially, both the Nature Conservancy and Ecotrust were instrumental to the alliance's operations; they "lent the alliance considerable assets ranging from foundation dollars and land holdings to a certain degree of credibility" (Hollander 1995c, 8). In recent years, however, the alliance gradually has weaned itself from these two groups by gaining the trust of the communities it was established to serve and developing financial resources of its own (Hollander 1995c; personal interviews, 1998; Willapa Alliance, 1995, 2).

New Governance Institutions: Trying to Turn Desperation and Gridlock into Trust, Respect, and Results

A number of citizens in the Applegate, Henry's Fork, and Willapa areas accepted the challenge offered by reconciliation and the idea that if they could just get the institutions right, they would be better able to discover the common ground necessary for building and sustaining a new community. Ideally, the new institutions would be more capable of maintaining and improving the high quality of life befitting their special place and to which they had grown accustomed. The new institutions would also foster the kind of understanding and trust necessary for enabling citizens to again live together as neighbors despite any differences in livelihoods or ideology. At the same time, it was an opportunity to develop

not a project in the sense that it has an end, . . . [but] a *process* aimed at building relationships so that people and groups will be able to effectively participate in natural resource management decisions and projects. The group's hope is that the development and empowerment of informal local networks will make the formal partnership obsolete. . . . Members . . . feel that the process is not just about [our community], nor is it just a [natural resources] issue; . . . it has broad implications for relearning lost community and social skills. (Shipley 1996, 3)

In short, participants saw an opportunity to build social capital and foster the kinds of skills necessary for strengthening and sustaining the democratic self-governance capacities of their respective communities.

The Basic Form
In each of the three cases—the Willapa Alliance, the Applegate Partnership, and the Henry's Fork Watershed Council (HFWC)—citizens opted for intermediary institutions designed to reconnect society to existing government institutions for the sake of improving the governance of their "special places." Each effort seeks to give citizens across the board a direct stake in the coordination and administration of policy using a collaborative, consensus-based decision forum. Government agencies—state, local, and federal—are asked to share power by relinquishing a certain amount of control, but not legal authority. An integrated, comprehensive approach to policy issues is taken, through both an emphasis on an ecosystem management approach and a tripartite focus on environment, economy, and community. And, as strictly advisory bodies, the new governance arrangements rely on negotiation, broad-based representation of interests, self-generated information regarding watershed conditions, and persuasion (rather than mandates and coercion) to shape policymaking and problem solving. (See also chapter 3 for a detailed discussion of the common institutional form, particularly as it relates to accountability.)

Missions
The Applegate Partnership officially "went public" in February 1993 after several months of meetings in the homes of interested citizens and stakeholders. It officially embraces the motto of "Practice Trust – Them

Is Us" and encourages supporters to sport buttons that symbolically signal that there is no them, only us (Rolle 1997a, 613). The mission statement targets environmental sustainability and community stability as primary goals. It reads as follows: "The Applegate Partnership is a community-based project involving industry, conservation groups, natural resource agencies, and residents cooperating to encourage and facilitate the use of natural resource principles that promote ecosystem health and diversity. Through community involvement and education, this partnership supports management of all land within the watershed in a manner that sustains natural resources and that will, in turn, contribute to economic and community stability within the Applegate Valley" (Shipley 1996, 2). The first board of directors, an elected body, put the mission into practice by agreeing on three action strategies:
• Provide leadership in facilitating the use of natural resource principles that promote ecosystem health and natural diversity,
• Work with public land managers, private landowners, and community members to promote projects which demonstrate ecologically sound management practices within the watershed, and
• Seek support for these projects through community involvement. (Shipley 1996, 2)

Officially chartered as a watershed council by the State of Idaho in 1994, the HFWC's formal mission statement sets forth three broad goals and four related major duties. The goals include (1) to serve as a grassroots, community forum that uses a nonadversarial, consensus-based approach to problem solving, (2) to better appreciate the complex watershed relationships in the basin, to restore and enhance watershed resources where needed, and to maintain a sustainable watershed resource base for future generations, and (3) to respectfully cooperate and coordinate with one another and abide by federal, state, and local laws and regulations. Major duties for the HFWC involve cooperating in resource studies and planning that transcends jurisdictional boundaries; reviewing, critiquing, and prioritizing proposed watershed projects; identifying and coordinating funding for research, planning, and implementation and long-term monitoring programs; and serving as an educational resource for the Idaho state legislature and the general public on the Council's progress.

The Willapa Alliance opted for a much shorter mission statement: "The mission of the Willapa Alliance is to enhance the diversity, productivity,

and health of Willapa's unique environment, to promote sustainable economic development, and to expand the choices available to the people
who live here" (Willapa Alliance, 1995, 1). Less explicit, but still central
to the alliance's efforts, is the emphasis on local community members as
the primary locus for problem identification and resolution. Executive
Director Dan'l Markham writes that "the Willapa Alliance was founded
on the belief that locally based science, economic research, and education
can assist the people of Pacific County [i.e., the Willapa Basin] to promote
positive change. The founding board members sensed a need for an independent organization of primarily local people representing a range of
views and interests to assist our community to chart its own destiny. . . .
[The goal is] to empower people to implement effective solutions to their
own challenges and problems" (Willapa Alliance, 1996g, 2). There are
four primary program areas in the Alliance's strategic plan:

- Natural resource ecosystem management,
- Science and information
- Public education and involvement
- Sustainable economic development

Moreover, from the alliance's perspective, despite the critical role
played by Portland-based Ecotrust in founding the alliance, and the fact
that the alliance's mission clearly is inspired by the Ecotrust mission,[10]
there is nothing wrong with relying on "external agents of change . . . if
they assist local communities to accomplish goals set locally" (Willapa
Alliance 1996g, 2–3). The keys are Ecotrust's desire to "build on the
cultural and economic traditions of local communities, . . . to recognize
. . . that [local peoples'] goal [of] long term economic prosperity is inevitably bound up with our [Ecotrust's] goal, the conservation and restoration
of ecosystems, . . . [and to] act as catalyst and broker . . . in working
partnerships with local individuals and groups" (Cleveland 1997, 23–
24). Arthur Dye, Ecotrust vice president and representative on the alliance in the early 1990s, argues that "if you start out believing that it
really is possible for communities to take control of their own future and
that that's a good environmental strategy, then you start looking for the
best place to do it. . . . [Ecotrust] selected Willapa Bay for the leadership
strength and natural capital it saw in the communities here" (Hollander
1995g, 15).

Getting Involved: Who's Participating?
Participation in all three governance efforts is broad based, ranging across virtually the entire spectrum of interests in each area. For example, in the HFWC case, integral participants come from a broad variety of environmental organizations (e.g., Greater Yellowstone Coalition, Henry's Fork Foundation, Idaho Rivers United, Nature Conservancy), state- and federal-level-agencies concerned with the management of public lands and natural resources, and local administrative officials such as planning and zoning or weed control personnel. There are also farmers and irrigators, ranching interests, recreation interests focused on fishing, hunting (Rocky Mountain Elk Foundation), and off-road motor vehicle use (Blue Ribbon Commission), independent and university-based scientists, and others who "reside, recreate, make a living and/or have legal responsibilities in" the watershed (Johnson 1996, 1). In fact, sixty participating organizations and agencies are listed in official council documents, including a local college (Ricks College), Fremont County Commissioners, Northwest Policy Center (University of Washington), Idaho Farm Bureau, Wool Growers Association, and Shoshone Bannock Tribes, among others (see appendix B for the full list). There is also occasional participation by citizens associated with the Wise Use movement and by legislative staffers for elected representatives from the U.S. House and Senate (U.S. Representative Mike Crapo, R-ID, now U.S. Senator; Senator Larry Craig, R-ID; Senator Dirk Kempthorne, R-ID, now Governor of Idaho).

HFWC meetings are held once a month using an all-day format and typically draw forty to sixty people. In the fall of 1998 these meetings were changed to a bimonthly schedule on the consent of HFWC participants. Meetings start with three minutes of silence and then facilitators set the ground rules for participation and deliberation. They remind participants about such things as the importance of civility, respect for others' views, and the prohibition of personal attacks. Two community-based organizations, the HFF and the Fremont-Madison Irrigation District (FMID), are the official cofacilitators of HFWC meetings, with the leaders of the two organizations—Jan Brown of the HFF and Dale Swenson of FMID—serving as the main facilitators. Yet facilitation is a team concept for the Council; others within each of the two organizations have received facilitation training and committee deliberations are often facilitated by participants other than Brown or Swenson (e.g., Charlie Sperry of the HFF, or Ed Clark of the FMID). Beyond facilitating meetings, the facilitation team "is chartered to attend to the administrative and logistical needs of the Council, coordinate its public information activities and submit annual reports of its progress to the [Idaho] legislature [as required by law]" (Henry's Fork Watershed Council, 1998, 3).

By comparison, the Willapa Alliance has a formal board of directors and employs a professional staff composed of an executive director, a scientist to direct the Natural Resource Program, a communications and education coordinator, an executive assistant, and a secretary/receptionist. Twice-monthly board meetings are supplemented with anywhere from one to four public forums each year that regularly draw 100 to 300 people from across the community, including bureaucratic officials and elected officials.[11] The seventeen-member board represents a broad cross-section of community interests and typically includes officials from large timber companies (Weyerhaeuser; Hancock Timber; Rayonier), an Ecotrust representative, small-scale tree farmers, oyster growers, cranberry farmers, environmentalists, fishing interests, the Shoalwater Bay Indian Tribe, as well as locally owned recreation and tourism businesses. There are also usually several other unaffiliated community members. For example, in 1995 a physician and an author/historian sat on the board (Hollander 1995a, 3; Willapa Alliance, 1995). Seats are not reserved for specific types of interests; instead there is an informal agreement that when a member of a particular interest category leaves the board, a concerted attempt will be made to fill it with another community member from the same category. Potential new board members are nominated and voted on by current members.

The Applegate Partnership also consists of a board of directors, yet public meetings open to all are held far more frequently than in the other two regions. In the early years, meetings were held twice each week; since 1996 the meetings have been held once a week. Each meeting typically draws six to eight members of the board of directors, with overall attendance generally in the neighborhood of fifteen to twenty people for each meeting. Unlike the Willapa Alliance situation, seats on the eighteen-member board are allocated among nine major stakeholder groups, with each interest represented by two co-board members. The groups represented on the board are local colleges, manufacturing industries, environmentalists, agriculture, local tree farmers, mining and mineral issues, timber extraction (large) companies, unaffiliated citizens, and a local research organization, the Rogue Institute for Ecology and Economy (RIEE). Initially, federal and state government representatives from public bureaucracies served on the board as well. However, a reinterpretation of the 1972 Federal Advisory Committee Act by federal administrators ended this practice in 1994 (Sturtevant and Lange 1995). Nominations to the Partnership Board in each of the interest groupings, while based on affiliation, are based as much or more "on a willingness to work toward solutions, leave partisanship at home, put ecosystem health in front of private agendas, and have time to participate in meetings" (Shipley 1996, 2).

Further analyses of meeting-level participation data for the Applegate Partnership and HFWC cases are provided in chapters 4 and 5, respectively. These analyses test whether the "surface" diversity evident here holds up to more rigorous scrutiny. Unfortunately, meeting-level data is not available for the Willapa Alliance.

The Question of Accountability

The emergence of GREM poses a clear challenge to traditional notions of public management because it involves a consensual, devolved style of governance in which power is shared among public and private actors, and citizens are actively engaged on a par with government experts in decision processes. A key challenge comes in the area of democratic accountability. Can the new governance arrangements known as grassroots ecosystem management produce positive-sum, or broad, simultaneous accountability without detracting from obligations and duties to state and national interests? Or does improved accountability to local interests have to come at the expense of accountability to broader public interests, whether it is state and national interests, or future generations?

Chapters 3 through 6 investigate these questions using interview data, government documents, primary documents, and published accounts of the Henry's Fork, Applegate, and Willapa cases of GREM. More specifically, chapter 3 describes and summarizes just what the accountability framework looks like from the perspective of the participants, and how it works in these three exemplary cases. Chapters 4 through 6, on the other hand, focus harder on the performance element of the accountability equation and report on the kinds of policy results or outcomes being produced by each case. Individual outcomes are also assessed as to whether they measure up to the promise of broad-based accountability evident in the institutional structure, processes, and practices described in chapter 3.

3

Operationalizing Accountability in a Decentralized, Collaborative, Shared-Power World

Participants in the Henry's Fork Watershed Council (HFWC), the Applegate Partnership, and the Willapa Alliance are keenly aware that their attempts to reform the institutions responsible for making and implementing environmental policy are being closely watched. They know that many, if not most, observers are skeptical about their ability to meet the twin goals of expanded accountability and improved performance, especially on a consistent basis. They are aware of the scathing criticism incurred by the Quincy Library Group (QLG) and that most Americans equate voting and elections with the idea of democratic accountability. Yet they are also firm believers that their nascent efforts to govern are legitimate, that as citizens they have a responsibility to get involved in governance activities, and that the critics are wrong, not having invested either the time or the effort required to understand and appreciate GREM.

The empirical investigation into participants' perspectives on accountability, as well as their institutional choices, suggests that the critics may indeed be wrong when it comes to the question of accountability. On the basis of interview data, government documents, primary documents, and published accounts of the three cases of grassroots ecosystem management (GREM), what emerges is a framework for accountability that is complex and holistic in approach. At the most basic level, participants are embracing governance arrangements that entail commitment to a concept of democratic accountability that is broad and simultaneous. The three cases employ a variety of means for achieving broad-based, simultaneous accountability, some of which are rooted in tangible institutional structure, processes, and management practices. Other accountability mechanisms, however, draw on the power of generally agreed to, yet

intangible informal institutions such as participant norms, the enculturation of specific virtues, and a credible commitment to accountability by
leaders. Participants in GREM also give philosophical, as well as practical, assent to a definition of accountability that includes demonstrable,
on-the-ground results. They believe that the explicit focus on performance adds meaning to the deliberations and conclusions bound up with
the political process; by explicitly connecting promises with outcomes,
this focus captures the essence of the democratic process.

Institutional Structure

Although there is some variation in degree among the cases, the three
cases of GREM display an institutional structure made up of seven key
elements, all of which are related to accountability (see table 3.1). First,
there is a collaborative, nonhierarchical (horizontal) design. The collaborative, horizontal structure means that power, leadership, information,
and responsibility are shared among participants. For example, "though
[Jack] Shipley is recognized as the charismatic leader of the Applegate
Partnership, his style of leadership is such that leadership and responsibility are shared equally among participants. All people have equal access
to power, information, and action" (Rolle 1997a, 614). Shipley himself
agrees, and believes that "because there [is] no lead agency or any one
individual in charge, all participants c[an] participate as equals" (Shipley
1996, 4). By not assigning leader and subordinate roles or formally denoting power differentials among participants in any way, participants see
the institutional design as better able to encourage discussion and negotiation and thus increase responsiveness to a broad variety of interests and
views (personal interviews, 1998 and 1999). Moreover, the literature on
collaboration suggests, and many participants believe, that by having all
collectively produced information out in the open (rather than hoarded
for strategic advantage), there are increased opportunities for innovation
and successful problem solving (personal interviews, 1998 and 1999; see
also Weber 1998).

A second critical structural feature involves citizens directly in the decision making related to the ecosystem or watershed in question, whether
in terms of identifying problems, crafting solutions, or implementing and
enforcing decisions. From this perspective, citizens are treated as coequal

Table 3.1
Institutional structure

Source of accountability	How it works	HFWC	Applegate Partnership	Willapa Alliance
Collaborative, nonhierarchical (horizontal) design	• Increases responsiveness to broad variety of interests • Increases opportunities for innovation and coordination	Yes	Yes	Yes
Direct, participatory decision making	• Treats citizens as coequal with government representatives and experts • Increases accountability to community members	Yes	Yes	Yes
Role and position relative to others • Catalytic, coordinative core • Advisory character	• Improves governance performance • Places constraints on independent action	Yes Yes	Yes Yes	Yes Yes
Place based character	• Facilitates cooperation, problem solving, monitoring, and enforcement through emphasis on a common bond • Promotes norm of integrity and honesty in communication and action	Yes	Yes	Yes
Open-access design	• Encourages diversity and responsiveness to concerns of the many	Yes, but few meetings	Yes, frequent meetings	Yes, but more insular
Open, public information systems	• Eliminates or minimizes information asymmetries • Facilitates monitoring and evaluation of activities	Yes	Yes	Yes
Legal charter	• Enhances legitimacy	Yes	Partial	No

to government agency representatives and experts; the concept of expertise is expanded beyond scientific, bureaucratic, and organized-interest expertise to include technical expertise in the community and citizen generalists with a "community" perspective (see also Scott 1998, 309–341). For example, citizen input concerning the quality of community life, or other values the community holds dear, is treated as a legitimate contribution on a par with technical recommendations by scientific experts. A member of a federal bureaucracy explains: "The HFWC increases our direct contact with citizens; . . . it is a bridge builder that helps us explain our decisions in a give-and-take format, and pass along new information about developments in the watershed. . . . [It] also serves as a refreshing . . . forum for new ideas and potential solutions" (personal interview, 1998). A major consequence of direct citizen involvement is an increase in accountability to participants from the community. Other consequences, according to GREM participants as varied as federal officials, independent scientists, and environmentalists, include new ideas and more comprehensive information about on-the-ground conditions that directly translate into improved governance performance (personal interviews, 1998 and 1999; see also Ostrom and Schlager 1997, 144–145).

A third structural characteristic concerns the role played by, and the position of, GREM arrangements relative to public agencies and private interests engaged in managing ecosystem resources. The new governance efforts are catalytic, coordinative cores that offer additional opportunities to decrease program redundancies, while also providing a one-stop communication forum for the many stakeholders in the area. In this regard, GREM participants say they are less interested in turf protection and in claiming sole responsibility for making the surrounding area cleaner or for preserving critical habitat, than in making sure key goals are achieved. As but one example of this mentality, members of the Applegate Partnership consider the proliferation of additional community groups focused on environmental sustainability and community health—primary goals of the partnership—something to be welcomed because they add to the critical mass focusing on sustainability (personal interviews, 1999).[1] In short, the new community-based institution does not have to actually "do" every project itself. Instead, participants seek either to coordinate and catalyze others' resources in support of the primary mission, or to encourage other like-minded efforts in the expectation of improved, more efficient performance.

At the same time, the three efforts are advisory in character and therefore are subordinate to public agencies and to state and federal laws, more generally. Nothing in the missions or legal charters of the Applegate Partnership, the HFWC, or the Willapa Alliance presupposes any right to circumvent or otherwise avoid obligations imposed by existing state and federal laws. Nor do any of these efforts have any legal authority to force private landowners into implementing programs or policies against their wishes. Instead, participants bear the burden of proof in convincing or persuading major stakeholders, whether public or private, to undertake actions to benefit the local community and environment. Thus, while the potential for efficiency benefits positively affects the performance side of the accountability equation, the subordinate position makes more likely the maintenance of accountability to state, national, and private (propertied) interests by placing significant constraints on independent action.

Fourth, the cases of GREM are place based—they are grounded in the local, nature-dependent communities. All three efforts studied here pay homage to the idea that their "place" is special and deserving of preservation. Participants from across the spectrum of private interests and public agencies agree with Kemmis (1990, 78) that place is a key catalyst for self-governance. This is because it helps to mobilize citizens to care enough to participate in the act of governing "their" place by reminding community members of what they share—the direct, tangible connection to, and reliance on the natural landscape (personal interviews, 1998 and 1999). More specifically, when taken together with the overarching mission of "environment, economy, and community" (see below), there is a clear sense among the people involved with these three cases of GREM that place can and does inculcate a sense of duty to the broader community. Sturtevant and Lange's (1995) study of the Applegate Partnership takes this a step further. They find that the "strong attachment to place" leads community members to "agree to put their interests, . . . and [their] sense of duty to represent . . . a particular perspective, . . . aside in the interest of the collective and [the] ecosystem" (10).[2] In addition, place factors into cooperation and problem solving; keeping the project and scope *locally focused* can help facilitate agreement between diverse interests (Shipley 1996, 4).[3] Observes Jim Neal, an aerial logger and cofounder of the Applegate Partnership in Oregon, "abstraction equals death for partnership, but once you . . . talk about a definable piece of land, you can get beyond philosophy . . . you can agree on what is acceptable and what is not" (Shipley 1996, 4). Finally, a number of participants note that the relatively close geographic proximity of neighbors facilitates both monitoring and enforcement, while also promoting the norms of integrity and honesty in communication and action (it is relatively easy for others to know about your behavior) (personal interviews, 1998).

The fifth structural element in the accountability framework is the open-access design found in these three cases of GREM. Generally speaking, all are welcome at meetings, including those who may wish to monitor and report on GREM activities to people outside ecosystem boundaries. The Applegate and Henry's Fork cases set the standard by enthusiastically welcoming outsiders and people from across the stakeholder spectrum to sit in on and participate in their proceedings. The Willapa Alliance, on the other hand, while clearly open to a broad crosssection of interests within the community, is more insular and therefore less comfortable with outsiders, either as observers or as full participants.[4] Most alliance meetings involve only board members and the executive staff, while mass public forums on specific issues (i.e., fish recovery strategy) occur anywhere from two to four times per year and typically involve 100 to 300 community members, representatives from organized interests, agency personnel, and elected officials (Willapa Alliance and Pacific County Economic Development Council, 1997). A local government official explains: "They don't have as many public meetings as the county does. But I remember when I first went to one of their public forums. I couldn't believe they had 200 people sitting there in the audience, many of them involved in the discussion. I mean we're lucky if we get 3 or 4 people to come to [the county's] public hearings" (personal interview, 1998).

Each of the three efforts practices government in the sunshine by voluntarily endorsing the community's right to know about its proceedings, decisions, and projects. At the most basic level, notices of meetings are publicly advertised in local media and with flyers posted throughout each community. Meeting notices and minutes of previous meetings are also mailed to participant-membership lists consisting of GREM participants as well as nonparticipants who have requested such notification. The HFWC, through the Henry's Fork Foundation, a key member of the council, also maintains a public library and geographic information system (GIS) database in their main office in Ashton, Idaho. In addition, the HFWC conducts public field trips and practices community outreach through schools as well as through regional and national watershed conferences (Weber 1999b). As a result, anyone who has a concern about HFWC activities, or perhaps simply wants to know more about the state of the watershed, has several means for doing so. The Applegate Partnership publishes a bimonthly community newspaper—the *Applegator*—that not only informs the community of partnership decisions and projects, but also offers a regular editorial forum for dissenters to the partnership as well as more general community news. In addition, the *Applegator* offers tips on such things as how to identify and protect flora and fauna important to ecosystem health, or how to cope with the periodic flooding of the Applegate River.

In combination with direct, participatory decision making, open access is viewed as not only increasing accountability to community members and any others who choose to participate, but also as encouraging diversity and responsiveness to the concerns of the many rather than the few. Moreover, it is seen as a way to foster direct, immediate accountability to the unfiltered concerns and preferences of citizens—the public interest as expressed by citizens—rather than an accountability grounded primarily in interpretations of the public interest by bureaucratic experts.

Sixth, collecting information and improving databases, whether scientific, cultural, and so on, is fundamental to the HFWC, Applegate Partnership, and Willapa Alliance (Henry's Fork Watershed Council, 1996; Klinkenborg 1995; Little 1997; Rolle 1997a; Shipley 1996; Willapa Alliance, 1995). The open, public character of decision making and information systems helps to level the playing field among interested parties by eliminating or minimizing information asymmetries in favor of one group or another (e.g., bureaucratic experts). As such, accessible information systems are viewed by participants as ways to enhance enforcement capacity and administrative legitimacy (authority). Interested citizens as well as elected and bureaucratic officials at all levels of government can access newly developed economic, social, and ecological databases to measure progress toward publicly stated goals and to assess the degree of (non)compliance themselves (personal interviews, 1998). For example, the Applegate Partnership, in collaboration with John Mairs, professor of geography at Southern Oregon University, has merged the various geographic information systems (GIS) and other analytical systems of the USFS and BLM related to the Applegate region. The Willapa Alliance has developed the Willapa Geographic Information System (WGIS) in cooperation with Pacific GIS and Ecotrust. The alliance argues that it "is the most comprehensive watershed-wide geographic information data base available to any rural community . . . and . . . [it is] one of the most powerful scientific tools available to facilitate conflict resolution and build consensus" (Willapa Alliance, 1995, 8). (See also the related discussion on monitoring and oversight in the "Institutional Processes" section below.)

Seventh, two of the three cases enjoy some level of formal endorsement by state government. In 1993, the Idaho state legislature passed a law encouraging advisory watershed councils in all state watersheds. The action allowed the HFWC to become state sponsored. The Applegate Partnership as a whole, on the other hand, is not formally sponsored by the state, but the Applegate River Watershed Council, a key committee considered to be the official implementation arm of the partnership, is. The official imprimatur of the state enhances formal legitimacy and establishes an accountability relationship with the state government and, by extension, with all citizens of Idaho and Oregon, respectively.

Institutional Processes

There are a number of institutional processes connected to the accountability equation that are evident in the decision making dynamic found in the Henry's Fork, Applegate, and Willapa cases (see table 3.2). The combination of joint deliberation and negotiation (rather than administrative fiat) with repeated interaction over time in each of the three GREM arrangements affects accountability in several ways. First, when coupled with the informal norms and institutions component, the experience of reasoning and deciding together in a civil fashion *over time* is thought to broaden individuals' horizons by exposing participants to others' views/situations, break down stereotypes, and facilitate additional

trust. The increased willingness to view issues from an "enlightened" perspective and to trust others encourages cooperative practices that build
community and increase the capacity for collective action on behalf of
adjacent ecosystems and communities.[5] Second, joint deliberation and repeated interaction, together with the open-access design and the public
character of information systems, level the playing field by improving or,
in some cases, equalizing access to information as well as to the rationales
behind governance decisions (minimizing information asymmetries that
may favor certain actors). Third, these elements of institutional process
encourage a systematic, comprehensive examination of evidence and values, improve the likelihood that all potential alternatives are explored,
facilitate the bargaining required for compromise solutions, and enhance
chances for project success by giving participants a bona fide stake in
outcomes (creates ownership in outcomes). Taken together, they directly
affect the performance element of accountability by increasing the likelihood that solutions will be properly "matched" to problems and carried
out effectively. Fourth, working together in a deliberative forum predicated on repeat interaction in a relatively small community tends to foster
the norms of integrity and honesty in communication and action. Purposely misrepresenting a position to gain advantage is high-risk behavior.
Any gains are likely to be short term, given the ability of others to personally check the credibility of stories and to impose sanctions of a social
and/or financial nature (e.g., less business from your neighbors) if these
stories are found to be untrue. At the same time, for those who have a
genuine, long-term stake in the area, it makes little sense to prevaricate
given the almost certain loss of influence in future group deliberations.

GREM also relies on a consensus decision rule, or a super-super-
majority, as opposed to majority rule (see table 3.2). This is seen as a
mechanism for improving the legitimacy of decisions by ensuring broad
agreement prior to programmatic action, while also increasing the likelihood of implementation success by lowering resistance and engendering
self-enforcement. Further, although winning the day does not require
unanimous agreement (100 percent consensus), the consensus rule, as
practiced in these three cases, appears to provide a formidable bulwark
against the abuse of minority rights. In each case, consensus, or lack
thereof, is determined by the general sense of the discussion rather than
by a formal vote. Once this point has been reached, someone summarizes

Table 3.2
Institutional processes

Source of accountability	How it works	HFWC	Applegate Partnership	Willapa Alliance
Joint deliberation and negotiation, and repeated interaction	• Encourages cooperative practices that build community and increase the capacity for collective action • Minimizes information asymmetries • Facilitates systematic, comprehensive examination of evidence and values • Increases likelihood that all potential alternatives are explored • Facilitates bargaining necessary to compromise solutions • Enhances chances for project success by giving participants a bona fide stake in outcomes	Yes	Yes	Yes
Consensus decision rule	• Increases legitimacy through broad agreement • Lowers implementation resistance and engenders self-enforcement • Respects minority rights	Yes	Yes	Yes
Standard procedure for major decisions	• Provides a checking mechanism against outcomes serving a narrow set of interests	Yes	No	No
Community-building exercises	• Fosters concern for, and obligation to, collective interests • Promotes civility and respect for others • Stresses equality of participants	Yes	No	No
Oversight/monitoring processes	• Creates additional assurance that programmatic goals are met and financial integrity occurs • Improves governance performance	Yes	Yes	Yes

• "Paper" review, physical inspections, evaluations by outside organizations, enhancement of citizen capacity	• Increases transparency		
• Community and ecosystem indicators		No	In process	Yes
Enforcement process	• Engenders self-enforcement dynamic	Yes	Yes	Yes

Table 3.3
Watershed Integrity Review and Evaluation (WIRE)

---

1. Watershed perspective: Does the project employ or reflect a total watershed perspective?
2. Credibility: Is the project based on credible research or scientific data?
3. Problem and solution: Does the project clearly identify the resource problems and propose workable solutions that consider the relevant resources?
4. Water supply: Does the project demonstrate an understanding of water supply?
5. Project management: Does project management employ accepted or innovative practices, set realistic time frames for their implementation, and employ an effective monitoring plan?
6. Sustainability: Does the project emphasize sustainable ecosystems?
7. Social and cultural: Does the project sufficiently address the watershed's social and cultural concerns?
8. Economy: Does the project promote economic diversity within the watershed and help sustain a healthy economic base?
9. Cooperation and coordination: Does the project maximize cooperation among all parties and demonstrate sufficient coordination among appropriate groups or agencies?
10. Legality: Is the project lawful and respectful of agencies' legal responsibilities?

---

the position of the group and suggests that consensus has been reached. If there is no opposition, the matter is considered concluded and the group moves on to the next decision. Generally, vociferous opposition by one or a handful of individuals at the suggested point of consensus is enough to extend the discussion until the points in dispute have been resolved to the dissenter's satisfaction, or to table the matter until a later date. Common objections that lead to tabling a proposal include the desire for additional information, and a concern that a stakeholder representing an important segment of the community needs to be present and part of the approval process. Participants also report that ideological or value-based objections are usually enough to prevent consensus from occurring.

Only one of the efforts—the Henry's Fork Watershed Council—has gone so far as to adopt a formalized process for major decisions. The HFWC uses the Watershed Integrity Review and Evaluation (WIRE) process, a formal procedural device that serves as a checking mechanism against outcomes serving a narrow set of interests. The WIRE process involves a series of steps (see table 3.3; see also appendix C for a sample

WIRE criteria checklist document) reminding participants that solutions must not only respect existing laws and agency mandates,[6] but must also reflect a total watershed perspective, rely on credible scientific data, emphasize ecosystem sustainability, and address the social and cultural concerns of community members, among other things. Initial review of the WIRE criteria occurs in a committee-based forum or discussion format. More specifically, participants break out into three committees—the technical or "hard" and "natural" sciences committee, the citizens' committee, and the agency roundtable. Members and tasks of each committee are specified in table 3.4. Once committee deliberations are completed, decision making is again managed under the purview of the full group. Projects receiving endorsement of the Council through the WIRE process may seek assistance, political support or interagency cooperation in their implementation (Henry's Fork Watershed Council, 1998, 3). Moreover, once a proposal garners consensus support, a subcommittee is formed for implementation purposes (e.g., Cutthroat Trout subcommittee; Water Quality subcommittee; Sheridan Creek subcommittee).

The Applegate Partnership and the Willapa Alliance, by contrast, adopt a broader *community visioning* approach for ensuring that decisions accord with their missions (personal interviews, 1998 and 1999). In both cases, community includes government officials with formal responsibility for public land and natural resources. The Applegate Partnership "involve[s] local communities in a visioning process of the Applegate watershed, whereby the partnership encourages the participation of individuals and agencies in developing a range of desired future conditions for the watershed: a community image of what the area should look like and how it should function. *Specific projects are then judged by how well they respond to the community-developed goal of forest health and community economic stability*" (Shipley 1996, 2; emphasis added). Most of the visioning for the Willapa Alliance, on the other hand, takes place internally, among the board of directors and the professional staff. Yet the large-public-forum format is an essential aid in the overall visioning process (personal interview, 1998).

Further, the HFWC is the only one of the three cases to start and end each all-day meeting with half-hour community-building exercises. The process is designed to focus attention on everyone's connection to place by emphasizing common ground and a shared sense of community.[7]

Table 3.4
Committees in action: Henry's Fork Watershed Council

Committee	Representation	Tasks
Citizen's Advisory Group	30-plus commodity, community, and environmental interests	• To review action proposals as well as agency proposals and plans. Are they relevant to local needs? Are all interests treated equitably? (see also WIRE process discussion below) • To forge new relationships/common bonds by offering a deliberative forum where watershed residents have a chance to get acquainted and establish trust where little has existed, thus building a sense of community and reducing the potential for agencies to play off conflicting interests against one another • To build credibility for the idea that the best course for the watershed emerges from neighbors who care about their common welfare
Agency Roundtable	20-plus federal, state, local, and tribal entities	• To help align existing policies with watershed resource realities and current public needs • To help ensure that Council decisions comply with the legal requirements of agency mandates and missions • To facilitate the practice of a "bottom-up" approach to natural resource issues by the agencies advocating such an approach
Technical Team	Independent, university, and agency scientists	• To serve as natural resource specialists, helping to ensure that Council decisions are based on sound science • To coordinate research and knowledge of existing research to minimize the potential for duplication • To integrate research results into Council discussions, decisions, and projects • To identify needed research and to monitor ongoing research

Sources: Henry's Fork Watershed Council, 1998, 2–3; Brown 1996b, 4.

During these periods there are no formal expectations as to subject matter; anyone can speak on any issue. A single ground rule governs interaction: personal attacks are prohibited. Participants sit in a circle and communicate personal stories (e.g., celebrating a child's achievement at school, notifying others of a neighbor's illness, telling about vacation experiences), or voice concerns on matters relevant to themselves or to the community. It also is common practice to recognize and thank others in the group or community at large for their efforts on behalf of the Watershed Council. Community building thus fosters concern for, and obligation to, collective interests, while also purposely promoting civility, respect for others' views, and the notion that all participants are equal.

The Applegate Partnership, for its part, does not conduct formal community-building sessions at the beginning and end of meetings. They do, however, occasionally gather for potluck dinners for the same general purpose (personal interviews, 1999). During meetings, community building is fostered through a focus on building "trust and respect among participants" (Shipley 1996, 2) because "it's an essential step in getting people to communicate what they want or what they think should be done" (personal interview, 1999).

A series of formal and informal oversight and monitoring processes reinforce the direct, almost immediate character of accountability found in these three cases of GREM. Like a more traditional administrative arrangement, the oversight and monitoring features are designed to create assurance that programmatic goals are met, financial integrity occurs, and institutional arrangements deliver on promises. Three mechanisms are of primary importance. There is a systematic review and discussion of projects once they are completed in an attempt to assess progress toward primary goals and to glean lessons for the future. For example, the HFWC sponsors an annual "State of the Watershed" Conference each fall to monitor the progress of Council-endorsed projects and to present research and monitoring results (Henry's Fork Watershed Council, 1998, 3). In many cases, such "paper" reviews are supplemented with physical inspections of, for example, the landscape or stream restoration project, or the timber cut in question (see the discussion of inspections as part of the ecosystem management approach below).

In addition, all three cases are subject to a variety of annual evaluations by outside organizations, including philanthropic foundations that provide temporary funding in support of objectives. These funding organizations require annual audits and, occasionally, outside reviews of activities by professional evaluators to ensure correspondence between stated goals and activities. Failure to satisfy external financiers means that funding will dry up, because in contrast to the situation with traditional public bureaucracy, there is no guarantee of future funding in the absence of demonstrable success in achieving goals. In the particular case of the HFWC, the outside review process includes a mandated report to the Idaho state legislature every year detailing its activities and progress (or lack thereof).

Moreover, participants are convinced that several structural features—direct citizen participation, open access design, and place-based character—extend oversight capacity by creating a cadre of skilled, informed citizens more capable of monitoring and evaluating public policy decisions and outcomes (personal interviews, 1998 and 1999). Thus, instead of relying solely on formal, often-sporadic inspections or hearings conducted by superiors (e.g., congressional oversight of bureaucracy), this is closer to the idea of a continuous monitoring process. The belief is that the creation of dozens or even hundreds of potential community-based, citizen monitors increases the chance that someone with the skill and interest will stop by, check up on a project's progress, and report what they have found to others. The extended monitoring capacity creates added incentives to keep projects on track and consonant with stated goals (personal interviews, 1998 and 1999; see also Tocqueville [1835] 1956). The Applegate Partnership has taken this one step further by successfully soliciting training assistance for citizens from the University of Oregon in the areas of water quality monitoring and forest health (personal interviews, 1999).

The sensitivity to making sure that proceedings and decisions are open and accessible to community residents as well as to any interested parties from outside the area has led the Applegate Partnership and the Willapa Alliance toward developing additional capacity to monitor and measure results. A more general result has been an overall enhancement of their oversight and decision making capabilities. Participants in each effort are convinced that strengthening accountability requires the development of *community indicators* capable of measuring changes in various elements of the environment, economy, and community mission (personal interviews, 1998 and 1999; see also Hannum-Buffington 2000; Sawicki and Flynn 1996; Sierra Business Council, 1999). Part of the embrace is due to past experiences:

Years of intensive forest management activities on public and private lands have taken place with neither adequate information about current conditions nor a mechanism for evaluating results.[8] As a result, we are faced with a variety of unintended consequences including loss of biological diversity, dramatically increased risks of fire and an overall decline in forest health. Similarly, community development activities, particularly those focused on rural communities, have taken place without clear assessment of current conditions, community-generated vision or systematic evaluation of the effectiveness of intervention (Lead Partnership Group, 1996, 35).

Yet participants also recognize the positive value of an effective monitoring system:

• It assists communities in determining whether they are making progress toward well-being and related self-defined goals.

• It assists communities in their evaluation of assumptions, both internal and external, about a community's characteristics, behaviors, and relationships.

• It informs institutions and organizations—business, government, and special interests—about their influence and impact in local areas. (Lead Partnership Group, 1996, 35; personal interviews, 1999)

A federal official engaged with the Applegate Partnership argues that to the extent that there is an effective monitoring and measurement system, it is "a way to increase the credibility of our efforts by making them more transparent, to show people that we are truly committed to the environment, economy, *and* community, that we're more than symbols and rhetoric, and that we're not controlled by special [private] interests" (personal interview, 1999).

Yet while the Willapa Alliance has embraced community indicators from the very beginning, the Applegate Partnership indicators project is still in the early stages of development (see chapter 5). Nonetheless, once completed (if completed), the partnership plans to formally incorporate the information into their planning and decision making processes. This kind of detailed information on the community permits greater scrutiny by participants and nonparticipants, and creates a richer, more accurate database regarding local conditions. Participants feel that the "better" information improves governance performance by helping them to not only make better decisions in the first place, but to fine-tune and adapt programs over time as conditions change.

Finally, given the advisory status of the new governance arrangements and because compliance is largely voluntary,[9] enforcement is a matter of collective responsibility in which self-enforcement (in which citizens agree with and willingly enforce policy decisions) and community norms (the informal rewards and sanctions that accompany community members' choices) play critical roles.[10] The self-enforcement dynamic draws heavily on a combination of self-interest (individual benefits), enlightened self-interest (the prospect of both individual and collective benefits), and the norm of trust as obligation. The norm of trust as obligation facilitates self-enforcement because, by definition, the practice of trust as obligation by participants means that they cognitively accept obligations to others and are willing to act to honor such obligations (Braithwaite 1998, 344). Moreover, as some observers of collaborative efforts note, the social pressure for individual performance to match public commitments in the collaborative arrangement tends to improve the quality of individual performance because "the individuals kn[ow] they [will] have an audience for their performance, an audience that [can] be appreciative but that [can] also be critical" (Bardach and Lesser 1996, 205). Or, as one participant in the HFWC puts it, "if you are acting on behalf of the Council, they [participants] watch you like a hawk" (personal interview, 1998).

Participants aware of potential noncompliance generally rely on social persuasion, reminding potential defectors of their collective obligations, voluntarily agreed to, and warning them of the possible consequences of their action (e.g., expulsion from the ecosystem management effort; loss of community "status"). The expectation is that these sanctions are severe enough that, over time, and in combination with the gains from collaboration and the disincentive provided by the high costs of the alternative—traditional "conflict-oriented" politics—outcomes will become self-enforcing. In addition, the open style of governance empowers more people and is conducive to enforcement because interested citizens can access information to measure HFWC progress toward publicly stated goals and assess the degree of (non)compliance themselves.

Norms, Virtue, and Committed Leadership

GREM advocates argue that this approach is much more than simply a set of formal rules governing the interaction of people with respect to place and that it is much more than the realignment of such rules to better reflect stakeholders' economic self-interest. Participants recognize that informal institutions are critical for achieving and sustaining governance performance and accountability (personal interviews, 1998 and 1999; see also Ensminger 1996; North 1990). Of chief importance are participant norms, the enculturation of virtue in participants, and a credible commitment to broad-based accountability by leaders. The dynamic generated by the informal institutions appears to approximate what Behn (2001) calls 360° accountability. The concept of 360° accountability acknowledges that a compact of mutual, collective responsibility, wherein participating individuals are accountable to everyone else, "true accountability for performance requires . . . a compact of mutual, collective responsibility [in which] . . . every individual [is] accountable to everyone else," including outsiders and/or those not present at a particular meeting, is required in order to achieve "true" accountability for performance (232).

Participant Norms

The Willapa Alliance, Applegate Partnership, and HFWC all rely on a set of informal participant norms, or behavioral expectations for all participants, as a primary source of accountability. The norms have evolved over time and are now part of an implicit bargain individuals strike prior to joining governance deliberations. Because the norms appear to be "well-crafted and well-diffused," especially in the HFWC and Willapa Alliance cases, they help regularize expectations regarding participant behavior by acting as substitutes for formal structural controls. Success here requires that group facilitators and individual participants readily and regularly enforce the norms when violations occur during deliberations, and that individual members regularly observe the norms as a part of their lives outside formal GREM meetings.

Examples of prevalent norms include civility and respect for others (and their positions), integrity and honesty in communication and action, and a commitment to the balanced "environment, economy, and community" mission and holistic, integrated approach to problem solving (see

Table 3.5
Participant norms

Participant norm	How it works
Inclusiveness	• All have a right to participate once they accept the norms governing participant behavior • Concensus building is reinforced by asking questions considering the preferences of "absent" interests
Civility/respect for others	• Each participant has equal worth and is afforded equal opportunity to influence decisions • Civil debate is stressed • Shared deliberation is encouraged (no monopoly of the discussion) • Violence, or the threat of it, is out of bounds
Integrity/honesty in communication and action	• Integrity is necessary for the success of the community-building goal and good-faith bargaining required to solve problems
Dual role (as community member and representative of particular interest)	• The process obliges participants to take a broader view of problems • Participants become committed to helping their neighbors solve community-based problems
Commitment to balanced mission and holistic approach	• Patterns of decisions benefiting only one segment of the community, or one element of the mission, are considered unacceptable
Trust as obligation	• Participants are required to follow through on public commitments • Trust is essential to program performance and the self-enforcement dynamic

table 3.5). For example, in each of the three efforts all participants have the right to speak and, in fact, are expected to contribute, if for no other reason than to signal their (dis)agreement with others' positions. But the character of the participation matters. Participants are expected to be respectful of others and sensitive to the right of others to speak, which means that "getting on your soapbox in a way that suggests close-mindedness" or otherwise dominating discussions by "rambling on for ten to twenty minutes at a time" are frowned on (personal interviews, 1999).

In addition, the Applegate Partnership and the HFWC more so than the Willapa Alliance strive to focus on people as members of place (e.g., the Applegate; the watershed), not on interest affiliation and/or formal positions and titles. Each effort believes that encouraging interaction on an individual-to-individual basis rather than as representatives from a particular organization facilitates more constructive discussion and deliberation, and ultimately contributes to problem solving and goal achievement (personal interviews, 1998 and 1999). Acceptance of a dual role as community member and representative of a particular interest obligates each participant to take a broader view of problems, to take a Pogo-inspired "them is us" perspective. Those involved in these efforts also see the "dual-role" norm as a way to commit participants, especially government agency personnel, to do whatever is necessary to help their neighbors solve community-based problems *within the rules provided by their legal mandates.*

There is also the norm of trust as obligation. The practice of trust as obligation means that participants cognitively accept obligations to others and are willing to act to honor such obligations, whether, for example, by actively helping the group implement decisions or by following through on public commitments, voluntarily negotiated and agreed to (Braithwaite 1998, 344). Trust as obligation is essential both to program performance and, as previously noted, to the voluntary, self-enforcement dynamic.

Finally, participants are expected to value inclusiveness and, by extension, diversity in all respects. Inclusiveness means not only that diverse interests are welcome, but also that people consider the preferences of others in absentia. In practice this means, as Su Rolle (1997a, 614) of the Applegate Partnership notes, that "when considering issues or projects, a common question . . . is, 'Who else needs to be at the table' "? Sometimes, as one environmentalist acknowledges, it means asking "what would so-and-so say about this issue if they were here today?" (personal interview, 1998). Yet precisely because these three cases of GREM choose to observe a series of additional behavioral norms and because they are not indifferent to the character of individual citizens (see "virtues" discussion below), the practice of inclusiveness explicitly links rights with responsibilities (see also Galston 1995, 38). This means that diversity and inclusiveness are not celebrated for their own sake; there are strings attached. That is why table 3.5 explains the norm of inclusiveness in terms of everyone having the right to participate *once they accept the norms governing participant behavior.*

To the extent that participant norms are clearly and consistently emphasized as integral to business and are evident to those with political and bureaucratic power at the state and federal levels, trust is likely to be promoted not only among the participants themselves, but in decentralized, collaborative, participative venues as viable alternative decision making forums. The increase in trust stems from the increase in certainty that such an arrangement will be democratically accountable—the participants will respect the rights of others, will do what they can to promote diversity in representation, will respect the legal mandates of government agencies, will follow through on program commitments, will operate in a public, open manner, and will make decisions with due attention paid to the accountability concerns of citizens and groups outside the region. It thus will be easier to for state and federal officials to grant additional latitude to identify, explore, and solve place-based problems, in some cases by simply agreeing not to interfere. Idaho State Senator Laird Noh (R), rancher, board member of the Idaho chapter of the Nature Conservancy, and chair of the Senate Committee on Environment and Natural Resources for eighteen of his twenty-one years in the Idaho legislature, values the HFWC because "it saves us [politicians] from a lot of headaches. . . . Communities work together to solve problems and forge common policy positions rather than coming as individuals with hundreds of different, unaggregated views to the legislator, who then must sort through the views and interpret what the community wants to the best of their ability" (personal interview, 1999). A Washington State senator (D) echoes Noh's sentiments: "The Willapa Alliance fills a niche in governance that needs to be filled. . . . Any time you do things closer to home, the better off you are and the alliance is about as close as you can get to the grassroots. They have a better understanding of local problems. . . . Added to this is the fact that they have a good reputation for getting things done and for having a diverse group of citizens involved. They are well respected by many state-level bureaucrats and legislators" (personal interview, 1998). Moreover, a growing literature in political science and public policy suggests that to the extent such a "strong fabric of trust" informs the relationships among participants and those outside GREM, performance benefits for governance efforts (more effective achievement of goals) are more likely.[11]

Of the three cases, the Applegate Partnership has had the greatest difficulty maintaining collective commitment to participant norms, with the result being a decline in trust among participants and in the capacity to reach consensus as well as an "increased frequency of expression that involves blaming, fault-finding, [and] attacking" (personal interviews, 1999; meeting minutes, February 3, 1999). The weakening of the collective commitment to norms is directly connected to the open access element of accountability and the unwillingness of a particular segment of the Applegate—environmental activists from the town of Williams, Oregon—to observe the norms either inside or outside of partnership meetings. The activists' rigid ideology, behavior during group deliberations, and embrace of certain tactics in an attempt to achieve their goals all violate participant norms in some way.

The Williams activists support a "deep green," or heavily ecocentric perspective ("environment first, second, and always") rather than the balanced mission statement of the Applegate Partnership. They also appear less interested in good-faith bargaining and the give-and-take of Partnership discussions and decisions than in ensuring that all decisions comply with their particular conception of "correct" policy. Nor is general progress toward their primary goals, which might entail some compromise, acceptable. Instead, it is an all-or-nothing strategy (meeting minutes, February 3, 1999). Further, the Williams activists do not always abide by the norms of civility and respect for others. Not only do they often interrupt others, they regularly challenge the veracity of agency participants' statements and engage in rambling, abstract philosophical discourses that may or may not address the point under discussion and that effectively prevent dialogue for fifteen to thirty minutes at a time (participant observation, May 12, 1999; personal interviews, 1999).

In addition, the Williams group has resorted to intimidation (e.g., threatening letters to the USFS) and violence outside of partnership meetings. A classic case was the summer 1998 "storming" and brief occupation of the USFS Ashland Ranger District offices as well as the Medford, Oregon, offices of BLM by several dozen protesters from Williams (with children among them), outfitted in guerrilla-style clothing, including face masks to conceal identities. USFS employees involved in the Ashland event describe it as a "wave of people storming the building . . . it was very confusing and frightening" (personal interviews, 1999 and 2000). Concern at the BLM's Medford offices was so great that officials expelled the visitors and locked down the building.[12] In response, USFS officials now lock down their buildings whenever Applegate Partnership meetings are held at USFS facilities, and, in at least one case, the concern was such that an armed sheriff's deputy was asked to be present (this occurred at the May 12, 1999, weekly partnership meeting). Likewise, the BLM conducted public forums in the Williams area as part of their efforts to make participation easier, yet repeated incidents of tire slashing convinced them to stop holding these meetings out of concern for the physical safety of their employees (personal interviews, 1999; participant observation, May 12, 1999).

Virtue

According to some, fewer opportunities for exercising the "primary political art" of deliberation, combined with the extended reach of government programs into daily lives, has proved problematic for the exercise of citizenship in the United States (see Landy 1993; Etzioni 1996). In the latter case, it is argued that citizenship skills have been weakened, dependency fostered, and accountability endangered to the extent that government, rather than individual citizens, structures and takes responsibility for the daily choices available to citizens. In the former instance, the overarching emphasis on "the rights citizens enjoy and on how policy can enable them to most fully enjoy and exercise those rights . . . [has had] a . . . corrosive impact [on citizenship]. Rights pertain to individuals, not collectivities. They tell people what they are entitled to, not what they owe. A full-fledged concern for citizenship would require policymakers to concern themselves as much with its obligations as with its privileges" (Landy 1993, 20). From this perspective, achieving policy outcomes accountable to a broad public interest requires something more than institutional and procedural checks and balances. It also requires a citizenry instilled with the kind of civic virtue and character that leads them "to sacrifice private pleasures for . . . public purpose[s] . . . [and] to care [enough] about the common good . . . [that they will] attach their own well-being to that of a wider community" (Landy 1993, 21; Sandel 1996, 319–325).

Precisely because governance arrangements that rely heavily on direct, iterative, ongoing participation by citizens require greater capacity for self-governance by citizen-participants, virtue becomes increasingly important to the accountability equation. Glendon (1995), Galston (1995), Sandel (1996, 6), Wilson (1985), and others find that certain character traits are likely to strengthen the capacity for self-governance and, by extension, accountability. Such nostrums strike most GREM participants and supporters as "plain old common sense" (personal interviews, 1998 and 1999). They recognize that successful governance using the GREM format is about more than "just the manipulation of incentives but also the formation of character" (Galston 1995, 38; personal interviews, 1998 and 1999). And they understand, or believe, that participant norms help moderate individual greed, selfishness, and ambition by teaching citizens how "to exercise their own freedoms responsibly, . . . to respect the liberties of their neighbors, . . . [to be] moderate in their demands, and [to be] able to discern and respect the rights of others" (Glendon 1995, 5; personal interviews, 1998 and 1999). In short, the people taking part in the Willapa, Applegate, and Henry's Fork governance efforts are in agreement with the observation by James Q. Wilson (1985) that making decisions that serve the public interest in the long term "depends on private virtue" (as quoted in Galston 1995, 38).

Virtues, or elements of individual character, identified by participants as vital to democratic accountability in decentralized, collaborative, participative arrangements using the holistic management approach (described below) include law-abidingness, civility,[13] a work ethic, the capacity to delay gratification, independence, entrepreneurialness, tolerance and inclusiveness, and honesty and integrity in communication and action. The expectation is that if individuals do not come to the deliberative forums already in possession of these virtues, the informal institutional framework of GREM will eventually inculcate such virtues. For example, law-abidingness as a virtue is a straightforward concept; citizens accept and respect current local, state, and federal laws as legitimate and comply accordingly. Independence bespeaks a willingness to care for and take responsibility for oneself, and to avoid becoming needlessly dependent on others (Galston 1995). A businessperson who belongs to the Willapa Alliance states: "Since we lack much of the more formalized structure that goes along with implementation in traditional . . . bureaucracies, we treasure people who can get things done without a great deal of prodding and babysitting. . . . These are people who understand that the burden or responsibility for success often rests with the individual who has made the public commitment [to shepherd a project through to completion, for example] in the first place" (personal interview, 1998).

The virtue of independence is closely coupled to that of entrepreneur-ialness in the minds of many participants in the Henry's Fork, Willapa, and Applegate cases (personal interviews, 1998 and 1999). Participants value entrepreneurial virtues such as imagination, self-initiative (self-starters), determination to succeed, and the willingness of participants to always be searching for innovative ways to solve problems. A federal official involved with the Applegate Partnership has found that because "getting collaborative projects off the ground can be frustrating; it re-quires a good mix of people who won't let a few obstacles like phenome-nally complex [federal] agency rules stop them dead in their tracks, and it requires what I call thinking outside the box, or thinking in between the rules in order to see all the possibilities" (personal interview, 1999). Another participant from the recreation sector testifies:

Sometimes the key to solving problems requires somebody to play the role of "innovation coordinator." What I mean by this is that some projects that citizens bring to us, and that we think will be beneficial to the watershed, have already been turned down for funding by a granting agency. But the problem that the project is designed to solve is still there and oftentimes the agency [in question] offers no further assistance. It doesn't fit for them so they're out of it; they stop trying to solve the problem . . . [and] they refuse to think entrepreneurially. That's our job. We innovate our way to success by coordinating and connecting private individuals with other agencies and resources . . . sometimes this means cob-bling together lots of little pieces to make the whole work. (personal interview, 1998)

Committed Leadership
While the Willapa Alliance has an executive director appointed by a board of directors, neither the Applegate Partnership nor the HFWC formally appoints or elects leaders. Instead, Applegate Partnership and HFWC leaders have come from the ranks of major stakeholding groups. In the Applegate, leadership is found in environmentalists Jack Shipley and Chris Bratt, and, in the first few years, Jim Neal from the timber industry and Brett KenCairn from the Rogue Institute for Ecology and Economy. Another key leader in the partnership until recently has been Su Rolle, a USFS employee who also served as BLM's liaison to the effort. In the HFWC, the two chief facilitators are leaders of the two major protagonists in the timeworn battle over natural resources in the Henry's Fork area: Jan Brown of the Henry's Fork Foundation (HFF), a conserva-

tion organization interested in maintaining and preserving the watershed, especially the world-famous fisheries of the area, and Dale Swenson, director of the Fremont-Madison Irrigation District (FMID) board, which represents the vast majority of farms in the watershed.

Leadership for accountability purposes in these decentralized, collaborative, participative institutions, however, relies on more than just formal sources of authority or power. It also requires leaders who are credibly committed to broad-based, simultaneous accountability. Such leaders champion broad-based accountability by reminding participants of the need to accept collective responsibility for putting it into practice and by engaging in behavior that leaves no doubt that this kind of accountability is of paramount importance to the proceedings. The behavior of leaders—or more specifically, reputation, rhetoric, and actions—weaves a discernible pattern that signals the degree of their commitment. As such, strong commitment to broad-based accountability by leaders necessarily strengthens the overall accountability framework. In practice, the HFWC and the Willapa Alliance have experienced greater success than the Applegate Partnership in establishing the committed-leadership component of the accountability framework.

From the perspective of other HFWC participants, there is no doubt that Jan Brown and Dale Swenson are clearly committed to the holistic environment, economy, and community mission of the council as well as to the inclusive collaborative process and ecosystem management approach. A farmer who is also a member of the Council notes: "Jan and Dale believe in balance rather than an environment-over-economy approach, or vice versa" (personal interview, 1998). Equally important, the personal credibility of the cofacilitators helps them to communicate the perceived value of the council design and processes to others. Each has earned a reputation for integrity and honesty, for always treating others with respect, and for having a clear commitment to and stake in the surrounding area. Moreover, each has primary responsibility for laying out the norm-based ground rules at the beginning of meetings, and each has a reputation for consistently enforcing participant norms (personal interviews, 1998).

The Willapa Alliance has enjoyed a similar leadership dynamic for most of its existence. Dan'l Markham, the executive director from 1994 to 1999, is a fourth-generation Willapa-area resident and a practicing Evangelical Christian minister. He is a passionate proponent of the tripartite Willapa Alliance mission who wants to preserve the pristine beauty and natural resources of his "place," and improve community capacity to manage place-specific policy problems so "the magic of the Willapa is kept intact for future generations" (personal interview, 1998). Much like Brown and Swenson of the HFWC, Markham is considered a skilled leader with strong connections "to the pulse of the community" and possessing "tremendous people skills" (personal interviews, 1998). And, according to a Native American participant in the alliance, Markham is "a person whom you can trust, . . . a straight shooter . . . [who] believes that open dialogue among the broader community, . . . having as many people as possible be part of the process, are absolute necessities if we're going to convince others that we're not only good problem solvers, but are accountable to boot" (personal interviews, 1998). Only time will tell if the new executive director of the alliance will demonstrate the same level of commitment to accountability as Markham.

The reputations, rhetoric, and actions of leaders of the Applegate Partnership have also displayed a strong commitment to broad-based accountability. An independent scientist-participant says other participants and outsiders alike typically identify Su Rolle, a federal government employee, along with Jack Shipley, as "the driving forces behind the whole thing. . . . They make it work, in part by being such strong cheerleaders for the idea that forests and rivers can be managed cooperatively by all the players. They're often the ones who insist that we make sure that a broad range of perspectives are heard and incorporated into final decisions" (personal interviews, 1999). Yet the current overall level of commitment in the Applegate Partnership is weaker than in the other two cases for several reasons.

First, three leaders with a strong commitment to broad-based accountability have left the Partnership without any obvious replacements. Jim Neal, owner of a local aerial (helicopter) logging company and one of several originators of the Partnership idea, left in 1997. Su Rolle retired from federal service and relinquished her partnership role at the end of 1999. Though she left in part to spend more time with her teenage daughter, there was also a "burnout" factor associated with the responsibilities of her informal leadership position (personal interviews, 1999 and 2000). Brett KenCairn took his skills, the lessons learned from his work at RIEE, and his participation in the Applegate Partnership to another organization pursuing similar collaborative, environment-and-economy solutions, the Grand Canyon Trust.[14]

Second, Chris Bratt is one of the strongest supporters of the Williams's area environmental activists and appears, at least to some participants, to be wavering in his commitment to the "environment, economy, and community" approach of the Applegate Partnership. In fact, the growing tension and mounting distrust among participants given the breakdown in norms has reduced some participants to tears on more than one occasion (personal interviews, 1999).

Third, Jack Shipley's involvement with Partnership matters has visibly diminished. By his own estimate, he now contributes eighteen to twenty hours per week on average as opposed to the fifty to sixty hours per week during the first five years of the partnership.

Ecosystem Management Approach

The people participating in the Applegate, Willapa, and Henry's Fork cases of GREM believe that a key part of the accountability equation centers on government performance—the ability of government to actually deliver on promises—and adopt a management approach with an explicit focus on performance (see table 3.6). The preferred institutions of earlier accountability models emphasize compliance with regulatory rules as the proxy equivalent of policy success—a cleaner, healthier, or protected environment, for example—and thus disconnect rules from policy results (Kettl 1983; Knott and Miller 1987; Weber 1998, 94–97). Environmentalist Michael Jackson of the QLG, for instance, challenges people to walk the spotted owl forests of Northern California and see what forest management policy shaped by the ESA has wrought—habitat so crowded with undergrowth that it is unsuitable for the owl (Christensen 1996). Or, as participants in the HFWC have discovered, government rules and practices have not stopped, and in some cases have created, severe environmental problems for the Henry's Fork of the Snake River— a world-class trout stream (Henry's Fork Watershed Council, 1996).

Table 3.6
Ecosystem management approach

Source of accountability	How it works	HFWC	Applegate Partnership	Willapa Alliance
Focus on results/performance	• Signals that performance is a high priority	Yes	Yes	Yes
Holistic, integrated approach	• Combats weaknesses of existing fragmented approach • Recognizes and wrestles with trade-offs • Recognizes interconnectivity and interdependence among resources, habitats, and people	Yes	Yes	Yes
Broad knowledge base	• Interdisciplinary approach is viewed as essential to broad-based accountability	Yes	Yes	Yes
Flexibility, adaptability, experimentation (adaptive management)	• Increases responsiveness to ecosystem problems • Facilitates learning • Reinforces the focus on performance	Yes	Yes	Yes
Proactive problem solving	• Focuses on preventing problems	Yes	Yes	Yes

Participants in these three cases therefore rebel against the established pattern by employing a results-oriented approach emphasizing on-the-ground ecosystem conditions as a primary basis for decision making and evaluation of policy success. People are asked to suspend preconceived ideas of the "right" way to achieve the ultimate goal of a healthy, functioning ecosystem.[15] Field inspections—walking tours—involving a cross-section of participants are often used to examine the physical condition of the landscape. The belief is that "walking the ground" reinforces the connection to "place" and an appreciation for the importance of nature to the entire community, while focusing discussion on the actual problems of the ecosystem in question in order to better match problems and solutions. Physical inspections also are seen as a way to develop a deeper understanding of how a particular management strategy helps or hurts the ecosystem, thus facilitating performance assessments and breaching the chasm between abstraction and reality that often accompanies policymaking exercises (Johnson 1997; Van de Wetering 1996).

Other management elements feed into the general concern for governance and policy performance. The new governance arrangements adopt a holistic, integrated approach to management, in terms of both the crosscutting "environment, economy, and community" mission statement and the comprehensive ecosystem orientation. The preference for a holistic, integrated approach is a direct response to the weaknesses of the traditional fragmented approach promoted by virtually every major federal environmental statute and supported by a bureaucracy specialized according to individual natural resources and management tasks (Clarke and McCool 1996; Davies 1990; Rabe 1986). By treating individual policy problems within the larger context of entire ecosystems or watersheds, humans included (i.e., community and economy), participants see themselves as trying to improve accountability by building into the management process explicit recognition of and responsibility for the trade-offs inherent in any form of policy administration. Because not every goal can be maximized in every single decision, individual decisions may favor nature over humanity, or vice versa, or may prioritize among environmental protection goals, for example, and favor some over others given their importance to ecosystem health. Yet, while trade-offs among goals are treated as inevitable, the philosophy argues that every choice be made with an eye toward ensuring that sustainability occurs and the diversity and integrity of the ecosystem are preserved. Put differently, unlike the dominant approach to resource development, which focuses on the individual "resource" (e.g., trees) and its ultimate economic (human) value, these alternative institutional arrangements attempt to see the forest as well as the trees (Weber 2000a, 243).

Participants also view the new management approach as better able to avoid the spillover, redundancy, and reverberation effects common to the specialized and single-issue approaches of existing environmental/natural resource bureaucracies. The ecosystem orientation is sensitive to the extant and potential stocks of natural resources for the area as a whole. And it is concerned with developing a better understanding of the relationship among various pollutants and natural resources (in order to calibrate values for trade-offs), as well as insight into whether solutions actually solve a particular ecological problem or merely shift its harmful effects elsewhere. In short, the management approach recognizes that a particular watershed's "web of life" is interconnected such that individual, fragmented decisions by stakeholders often affect the health and well-being of other resources, habitats, and people.

Another key component of the management strategy is the reliance on a broad knowledge base. Such an approach accepts that real-world problems typically do not fit nicely and neatly into the *singular* domains of traditional academic disciplines, nor are they amenable to analysis excluding social impacts. As a citizen (unaffiliated) who is involved with the HFWC observes, while "natural and hard sciences are the key to unlocking natural resource problems, they don't come with the necessary instructions regarding how to apply them in human settings" (personal interview, 1998). As a result, social sciences as well as nontechnical, community-based knowledge are valued along with physical and natural sciences (e.g., silviculture, biology, ecology, chemistry) and technical professional advice (e.g., engineering). In a very real sense, these collaborative arrangements argue that an interdisciplinary approach is an absolute prerequisite if broad-based democratic accountability is desired.

In addition, the management style is proactive rather than reactive. Instead of waiting for problems to arise and then correcting them, by which time it is often too late and/or much more costly to mitigate damages, ecosystem data is gathered and organized in order to identify problems and prevent them from arising in the first place.

The overt focus on policy results, the use of management by walking around, the interdisciplinary character of analysis, the limited understanding of ecosystem dynamics that plagues all natural resource management efforts, and the inherent uncertainties associated with a crosscutting, integrated approach to policymaking have led GREM participants to adopt an adaptive management style. Instead of locking in the "best" solution, or automatically employing the same administrative structure for every program, learning through experimentation is the rule. Specifically, deliberate long-term experimentation with humanity's interaction with ecological processes facilitates the process of learning what works and what does not (Lee 1993, 8). Ideally, adaptive management increases responsiveness to ecosystem problems by promoting the values of continual innovation and adaptation in response to changing conditions, problems, and degree of success (or failure) enjoyed by solutions. It also helps to create an administrative mindset that more readily accepts the demise of nonperforming programs, while increasing the likelihood that programs that have successfully solved a particular problem either will be replicated in the future as similar problems arise or discarded at the appropriate time.

Focusing on Outcomes: Making Sure That Promise is Followed by Performance

Saying that certain sources, or mechanisms, of accountability exist and explaining how each element works, however, says nothing about the degree of assurance that broad-based accountability will actually occur. It thus becomes necessary to examine the actual decisions and outcomes being produced by GREM governance arrangements.[16] The examination of outcomes, or choices—the final component of the new accountability framework—again brings to the fore the distinctiveness of the GREM accountability concept in terms of how each relates to two questions central to accountability scholarship—the questions of "for what" and "to whom." The synergism of the overall institutional dynamic reinforces the broad-based, simultaneous notion of accountability by fusing these two questions into a single whole. The fusion makes clear that participants object to a conceptualization of accountability, as reflected in policy results, that consistently favors one set of interests over another, or that satisfies one agency's "narrow" mandate at the expense of new spillover problems for (un)related policy areas. Is theory reflected in practice?

Moreover, given that there are several different levels of government in the U.S. system, a successful claim for broad-based, simultaneous accountability requires an additional test. To earn the label of broad-based, simultaneous accountability, outcomes need to demonstrate accountability across the existing levels of government. As a result, the outcomes from the Applegate Partnership, the HFWC, and Willapa Alliance will be mapped against four different levels or scales at which accountability might occur—the individual (micro), community (meso), state/region (mid-macro), and nation (macro). The criteria for assessing accountability are as follows:

• The diversity of representation in processes and individual outcomes
• The relationship of choices (outcomes) to existing laws, regulations, and agency programs
• The extent to which choices result in benefits for a broad array of interests
• The specific effect of outcomes on "individuals" within the community

The Diversity of Representation in Processes and Individual Outcomes
The new governance arrangements explicitly embrace an inclusive, pluralistic approach to participation. The desire is to have discussion, deliberation, and decision making engage the entire spectrum of interests and stakeholders for a particular place, including state, regional, and national interests. The concern for broad representation suggests that solving the "for what" question is necessarily subsidiary to, and inseparable from, the question of "for whom." Therefore, the distribution of participants matters. Who is participating? And, if the data are available, who are the core participants? In other words, which citizens, interests, agencies, and levels of government tend to display steady, significant levels of participation? The diversity of representation criteria applies to each GREM governance arrangement as a whole as well as to each individual outcome.

Yet, precisely because not all participants exert equal influence over final decisions, the test of representation is an important measure of accountability but is not definitive by itself. One can easily imagine a scenario wherein balanced representation occurs, yet because the final outcome represents the few at the expense of the many, claims of broad-based accountability are suspect. One can also imagine the opposite being true. Representation in the decision making process is dominated by one set of interests, yet outcomes reflect a clear concern for broad-based accountability. This is why additional criteria are needed to determine if outcomes meet the simultaneous, broad-based accountability standard.

The Relationship of Choices (Outcomes) to Existing Laws, Regulations, and Agency Programs   The focus on the relationship of choices to local, state, and federal mandates accepts the legitimacy of how established governing units have defined "for what." The criteria recognize that the choices of the new governance arrangements may either hamper or help existing units of government achieve predefined public goals as articulated in laws, regulations, and programs. In other cases there will be "no conflict" because the endeavor in question is staking out new policy ground that does not conflict with existing mandates and programs. This criterion is thus primarily concerned with whether outcomes weaken, preserve, or strengthen accountability to particular levels of government. Accountability is *strengthened*, for example, in cases where outcomes help government agencies get more "bang for the buck" through things like improved coordination of agency efforts, or the catalyzation of previously moribund or unavailable (e.g., private-sector) resources in the service of previously defined public goals. Strengthening also occurs when innovative means are crafted for achieving the same publicly defined goals at the same or less cost. In all three examples, accountability for what increases by virtue of the fact that bureaucracies either extend the reach of their programs/missions at the same cost, or achieve similar levels of performance at less cost. Similarly, accountability is *preserved* when outcomes either display clear support for, or no conflict with, laws and regulations already on the books. Accountability is *weakened* when outcomes conflict with existing laws, force the expenditure of additional moneys for fewer results, or shift operational control of a public program to individuals or select groups in the private sector.

The Extent to Which Choices Result in Benefits for a Broad Array of Interests   The rhetoric and missions of the three cases of GREM clearly promote positive-sum, or win-win-win, outcomes benefiting a broad array of interests, at least as it concerns the community or place-based level of analysis. As part of this focus on positive-sum outcomes, there is a new way of thinking about the "to whom" question of accountability, especially as it relates to the question of accountability "for what." Participants are willing to support decisions that satisfy the "for what" component of accountability as defined by duly elected and constitutionally appropriate authorities; they are legitimate, case closed. But participants do not want this to be the end of the discussion about the "to whom" question. Instead, they argue that there is a need for creative new methods and programs that expand accountability to a broader array of interests (to whom) *without* harming the original intent (for what) contained in laws and regulations. The idea is that "for what" cannot become a reality until and unless the answer to the "to whom" question satisfies a broad-based coalition of citizen-participants. Failure here leads to the default option of no decision. The central question for this criterion thus concerns how well participants' program choices and efforts actually satisfy the quest for positive-sum outcomes benefiting the environment, economy, and community.

The Specific Effect of Outcomes on "Individuals" within the Community
Outcomes affect "individuals" within each community that hosts one of the new governance arrangements. For example, accountability to individuals is strengthened when outcomes focus on a single individual's problem, promote a zero-sum outcome benefiting certain individuals' interests at the expense of others, or provide clear support for specific livelihoods and/or segments of the local economy. Accountability to individuals is also strengthened when a limited array of interests are involved in the decision process or when an outcome is sensitive to the unique characteristics of individual situations (a custom fit). Conversely, accountability to individuals weakens when outcomes provide less support for specific livelihoods and/or particular segments of the community, or promote a shift toward the idea of enlightened self-interest.

However, the ratings of weak versus strong hold different connotations at the individual level of accountability as compared to the other three collective levels of accountability. Because the purpose of the assessment is to determine whether the outcomes support a conclusion of broad-based accountability, a rating of "weak" accountability for the individual level holds the same meaning as a rating of "strong" for the community, state, and nation levels of accountability. In other words, both ratings indicate support for broad-based accountability. The converse is also true. A conclusion of "strong" accountability at the individual level is the equivalent of a "weak" rating for the other three levels (i.e., there is little support for a conclusion of broad-based accountability).

In any case, the key to assessing the degree of simultaneous, broad-based accountability for each outcome is to weigh the balance of the evidence from the various criteria. A rating of weakens, preserves, or strengthens for any one criterion at any one level is not very useful. Rather, given the many possible combinations of accountability criteria for each outcome, it makes more sense to picture the concept of simultaneous, broad-based accountability along a continuum from strong to weak. In general, the strongest case for a conclusion of simultaneous, broad-based accountability is when an outcome

• Includes representatives from two or more levels of government during the decision making process
• Exhibits balanced representation among the major stakeholders and the public agencies responsible for natural resources and land management
• Offers clear support for a positive-sum environment-and-economy outcome
• Extends the program reach of existing government programs at two or more levels of government
• Weakens accountability to individuals (e.g., less support for specific livelihoods and/or particular segments of the community)

The weakest case, on the other hand, is an outcome that does not include any government representatives, is dominated by a single set of interests, promotes a zero-sum choice, conflicts with existing law at two or more levels, and strengthens accountability to individuals. Moreover, it is important to recognize that an outcome that strengthens accountability at the individual level does not automatically place it toward the weak end of the continuum. Indeed, it is entirely possible to have an outcome strengthen accountability to an individual segment of the community (e.g., farmers) while simultaneously strengthening or preserving accountability to the community, state/region, and nation.

Chapters 4 through 6 review a total of thirty outcomes produced by the Applegate Partnership, HFWC, and Willapa Alliance. If the common criticisms are valid, there should be a large imbalance in the way accountability is apportioned among the levels. Most outcomes should end up strengthening accountability to the individual and/or community (meso) levels at the expense of the state/region and national levels. On the other hand, outcomes that meet the demands of a broad-based, simultaneous form of accountability will exhibit accountability to multiple interests and levels at the same time. Of equal importance, the outcomes suggest that these new governance arrangements, rather than only being about process and relationship building, are also about developing new capacities for problem solving that simultaneously promote environmental protection, economic concerns, and community sustainability.

4

The Applegate Partnership: "Practice Trust, Them Is Us"

When Jim Neal, of the Aerial Forest Management Foundation, and Jack Shipley, of the Headwaters environmental group, convened the initial meeting of sixty Applegate watershed community members in the fall of 1992, there was little more than a hope, a wish, and a prayer that their efforts might lead to something as enduring as the Applegate Partnership. The war of all against all dominated the landscape. Public land agencies and the experts therein were used to managing natural resources with little direct interaction and guidance from the community-based stake-holders whose lives and livelihoods were most heavily affected by agency decisions. Current management practices and institutions appeared to be poorly matched to a whole series of emergent wicked problems. Could anyone but a hopeless, naive idealist honestly expect that a collaborative effort centered on maintaining the long-term health of the watershed and stability of the local economy, and with a motto of "Practice Trust, Them Is Us," would still be alive ten years later?

More importantly, how well has the coalition of the unalike that is the Applegate Partnership delivered on the original promise to produce outcomes designed to placate a broad public interest? Does accountability stop at the borders of the Applegate watershed? Or does it extend to the regional and national levels? Are certain interests and policy outcomes typically favored at the expense of others? Does the Partnership have a genuine commitment to openness, diversity, and existing government mandates? How well do the Applegate Partnership's program choices and efforts satisfy the idea of broad-based accountability? Choices and efforts include projects directly managed by the Applegate Partnership, and the facilitation of new groups and processes that take as their starting point the broad-based environment, economy, and community mission of the partnership.

Relying on nine outcomes, or observations/cases,[1] the short answer is that it is not clear that the standard, often historically grounded criticisms and concerns about accountability are valid when applied to the outcomes produced by the Applegate Partnership. In a series of cases ranging from intervention into a state-mandated, yet highly fragmented county planning process to landscape-level forest planning, the active initiation of new, locally grounded economic markets, and efforts to reintegrate "neighbors" and privately held lands into a more environmentally sensitive, watershed-wide resource management plan, the evidence supports a verdict of broad accountability as the norm.

Diversity of Representation

The Applegate Partnership's board of directors is purposely designed to reflect the diversity of the major stakeholding groups in the watershed. The seats on the board are allocated among agricultural interests, environmentalists, unaffiliated citizens, timber extraction (large) companies, local tree farmers, local colleges, mining interests, manufacturing industries, and the Rogue Institute for Ecology and Economy (RIEE). Table 4.1 displays the distribution of participants for 205 Partnership meetings, or approximately 53 percent of all meetings held from January 1994 through July 1999.[2] A total of 597 people, the equivalent of 5.0 percent of the Applegate's total population, have attended at least one of the 205 meetings, with an average of 18 people in attendance at each meeting.[3] Of the 597 participants, only 460 or 77 percent are identifiable, hence the population used in this analysis equals 460.[4] Further, of the 137 unidentified participants, fully 91.2 percent (125) attended only one meeting. Of the remaining 12 unidentified participants, 8 attended two meetings, 3 attended three meetings, and 1 attended five meetings. Put differently, although the analysis contains only 77 percent of all participants ($n = 460$) for the five and one-half year period, it covers 95.7 percent of the total meeting visits.

The analysis in table 4.1 considers the distribution of participants for all 205 meetings and as a function of the frequency or intensity of attendance by participants. For all Applegate Partnership meetings, there are significant levels of participation among three key groups with major stakes in the watershed. Federal administrative agencies—organizations responsible for administering national mandates—register an average participation rate of 28.0 percent, the largest of any category. Environmentalists and commodity interests, primarily timber in this case, record relatively similar rates of participation, with 16.3 percent and 12.8 percent, respectively. Among other categories, unaffiliated or concerned citizens provide a strong presence with 20.4 percent of all participants, with scientists/academics registering a participation rate of almost 11 percent and state-level administrative officials at 5.9 percent. Conspicuous by their low participation rates are recreation interests, who record a participation rate of 2.2 percent (7 of the 460 participants), elected officials of all levels (0.2 percent combined), local administrative officials (2.5 percent), and the electronic and print media (0.8 percent).

Columns 2 through 4 in table 4.1 examine how the frequency or intensity of attendance by individual participants affects the balance of voices at partnership meetings. The measure of intensity is important because it tells us the makeup of the core groups of repeat players. Repeat players are participants who possess a higher degree of commitment to the mission, have an institutional memory to match, and are likely to exert greater influence over final decisions. Intensity of participation is measured in four categories—those attending 10 or more meetings ($n = 62$), 20 or more meetings ($n = 39$), 51 or more meetings ($n = 15$), and those exhibiting a high level of commitment by attending 103 or more meetings ($n = 8$).

Unsurprisingly, the committed core groups of repeat players are much smaller than the total number of partnership participants for the 205-meeting sample. Further, as the intensity of commitment to the new governance arrangements rises from the bare minimum (attendance at one meeting) to the extremely committed category—attendance at more than 50 percent of the meetings—there are several significant changes in the distribution pattern. One of the clearest lessons to come from the frequency analysis is the overall dominance of environmentalists. In all cases except one, environmentalists participate in the partnership at rates on average double that of commodity interests. In the most striking case of dominance—those who have attended 25 percent or more of all meetings—environmentalists comprise almost 50 percent of all participants.

Table 4.1
Distribution of participants in the Applegate Partnership, sample of 205 meetings from January 1994 through July 1999

Participant category*	Total attendance (205 meetings) n = 460 %	Attends 10+ (5%) meetings n = 62 %	Attends 20+ (10%) meetings n = 39 %	Attends 51+ (25%) meetings n = 15 %	Attends 103+ (50%) meetings n = 8 %
Environmentalists (Headwaters, Nature Conservancy, Sierra Club, National Audubon Society, American Forests)	16.3	30.7	33.3	46.6	37.5
Extractive/commodity interests (farmers, irrigators, ranchers, timber, local business and development interests)	12.8	14.5	18.0	20.0	37.5
Recreation interests (Oregon Trout; Rogue Flyfishers)	2.2	4.8	7.8	6.7	0.0
Federal administrative officials (BLM, USFS, National Marine Fisheries Service, U.S. Fish and Wildlife, EPA, Army Corps of Engineers, U.S. Department of the Interior)	28.0	25.8	17.9	13.3	12.5
State administrative officials (State Forestry Extension, Oregon Jobs Council, Departments of Agriculture, Forestry, Fish and Wildlife)	5.9	1.6	0.0	0.0	0.0
Local administrative officials	2.5	1.6	0.0	0.0	0.0
Federal legislative representatives	0.2	0.0	0.0	0.0	0.0
State legislative representatives	0.0	0.0	0.0	0.0	0.0
Local (elected) officials	0.0	0.0	0.0	0.0	0.0
Concerned citizens (no formal group affiliation)	20.4	11.3	12.8	6.7	12.5
Independent and university-based scientists	10.9	9.7	10.2	6.7	0.0
Media	0.8	0.0	0.0	0.0	0.0
Total attendance	100	100	100	100	100

* Participants representing two organizations—the Rogue Institute for Ecology and Economy (RIEE) and the Applegate River Watershed Council (ARWC)—with broad missions do not fit neatly into any of the primary categories. The RIEE (located in Ashland, Oregon, just outside the watershed), is heavily involved in the Applegate Partnership and holds down one of the board seats. Its thirteen participants are thus split between the local and regional categories, with six treated as locals and seven as regional. At the same time, the RIEE has a research-and-advocacy mission that considers ecology and economy inseparable (Johnson 1997). As a result, the thirteen participants are split among the following categories: three in the environmentalist category, three in the commodity, three as recreation, and four as academics. The ARWC, on the other hand, is an integral part of the Applegate Partnership and is located in the watershed. Therefore, its twelve participants are all treated as locals. Yet the ARWC also pursues a multifaceted mission grounded in environment, economy, and community. Therefore, three participants are assigned to the environmentalist category, three to commodity, three to recreation, and three to concerned citizens. A third organization, Southern Oregon University (SOU), does not fit neatly into the "location" categories. A state university located in Ashland, Oregon, SOU participants occasionally serve on the Applegate Partnership board in the local colleges' slot. Therefore, the four SOU participants are split equally between local and regional categories.

Table 4.2
Distribution of attendance by location, 205 meetings ($n = 460$)

Participant category	Local %	Regional/ state %	National %	Totals %
Environmentalists	12.4	2.8	1.1	16.3
Extractive/commodity interests	8.9	3.5	0.4	12.8
Recreation interests	1.1	1.1	0.0	2.2
Administrative officials/bureaucratic experts	2.5	5.9	28.0	36.4
Legislative representatives (elected or staff)	0.0	0.0	0.2	0.2
Concerned citizens	18.7	1.7	0.0	20.4
Independent and university-based scientists	2.2	6.1	2.6	10.9
Media	0.4	0.4	0.0	0.8
Total attendance	46.2	21.5	32.3	100

In the only case that belies this larger pattern, environmentalists represent 37.5 percent of all participants, exactly the same as commodity interests (see column 5 of table 4.1). Also of significance is the virtual disappearance of state-level bureaucratic experts (1.6 percent) in the first category (attends 5 percent or more of all meetings), accompanied by their complete disappearance from the remaining core groups. Federal-level administrative officials, for their part, display a steady decline in participation rates, from 28.0 percent of all participants for all meetings, to about half that rate when the frequency of participation exceeds 25 percent of all meetings. On the other hand, recreation interests exhibit a substantial increase in participation rates across the first three of the four core groups.

Another way to think about accountability using meeting attendance data is to explore the location of the individual participants. Where do Applegate Partnership participants reside? Are they locals? Do they come from the surrounding region (greater Oregon, northern California, Washington), or from other parts of the United States? Table 4.2 displays the results of this analysis. To no one's surprise, local participation dominates the participation statistics with a score of 46.2 percent. Roughly one out of every five participants represents the surrounding region, while one out of every three participants is from the national level. Within the "local" category, the unaffiliated citizens are the dominant group, with roughly 40 percent of all local participants. Environmentalists comprise the second largest group of locals with just fewer than 27 percent of the total, and commodity interests rank third with 19 percent. At the same time, the strong "national" participation reflects the land-ownership pattern in the Applegate watershed. Almost 87 percent of all national-level participants represent natural resource bureaucracies, with virtually all of these administrative officials coming from the USFS and the BLM, the two agencies that control 70 percent of the Applegate's land. Finally, as most would expect given the logistics of attending Applegate-based meetings, the vast majority (over 91 percent) of unaffiliated citizens are locals (but see the HFWC statistics in chapter 5 for a much different story).

In sum, unless we assume that state- and national-level participants are captured or somehow unduly influenced by local representatives in Applegate Partnership proceedings, the meeting-level participation data suggest there is little risk of local interests usurping or otherwise ignoring state and national interests. The weakest link in this regard—to the state level—still comprises 21.5 percent of all participants, although only one state-level bureaucrat has attended more than 10 meetings (in the 205-meeting sample). Moreover, the participation pattern evident in the partnership runs contrary to conventional wisdom in terms of which groups will dominate decentralized governance arrangements. In fact, it is the environmentalist contingent that clearly dominates proceedings, especially once frequency of attendance is taken into account.

Thus, it is plausible to argue that the Applegate Partnership case enhances accountability to individuals and a broad cross-section of the community, while simultaneously preserving a solid measure of accountability to the state/regional level and strong measure of accountability to the national level (see table 4.3). Of course, the outcome analysis does not rest on a single case. The possibility exists that meeting-level statistics mask a deeper political dynamic that may accord with the skeptics' fear of agency capture and dominance by local interests, particularly economic interests. If the skeptics are right, the pattern in the rest of the outcomes should overwhelm and belie the lessons derived from the meeting-level participation data.

Table 4.3
Linking Outcomes to Accountability: The Applegate Partnership

Outcomes	Individual (micro)	Community (meso)	State/region (mid-macro)	Nation (macro)
Diversity of Representation	• Strengthens/ open access, iterative design equals greater opportunities for individual citizens • Strengthens/ dominance by environmentalists	• Strong/a broad cross-section of community-based stakeholders	• Preserves, although it is a weaker case than the macro-level case (see table 4.2)	• Preserves (see table 4.2)

Staying True to Existing Mandates

The Applegate Partnership statement of purpose and written brochures have expressed allegiance to national and state laws from the beginning. An environmentalist remarks that "from day one, [the Partnership] has been a supplementary system of accountability intricately connected to and nested within the larger state and federal system of accountability" (personal interview, 1999; see also Applegate Partnership, 1996a). At the same time, there is a strong interest on the part of partnership participants in being able to facilitate more effective implementation of existing legal mandates by incorporating creative new methods of implementation. In turn, the expectation is that new methods of implementation will offer opportunities to expand accountability to a broader array of interests ("to whom") without harming the original intent ("for what") contained in laws and regulations. More colloquially, as a federal official explains, this means "thinking outside the box of regular or traditional program prescriptions" (personal interview, 1999).

In its most basic form, staying true to agency mandates is about how the Partnership facilitates the production of better information and, according to an environmentalist-participant, increases the legitimacy of agency decision making by "providing the federal agencies the public involvement component that historically they've never had" (personal interview, 1999). Yet, as a USFS official explains, the public involvement has not led to co–decision making for public lands; agency personnel are still the official decision makers as required by law. Instead, the partnership has become "an important source of information about what the community wants" (personal interview, 1999). Another federal official finds that "the diversity of participants and the discussion-oriented forum are such that we know we're getting probably as close to a full . . . cross-section of true community viewpoints as any public forum. This helps us make decisions better suited to the community, politically speaking, but without sacrificing the [policy] goals that Washington gives us [whether legislative or administrative in origin]" (personal interview, 1999).[5] Further, the partnership has been instrumental in coordinating the integration of BLM, USFS, and private-sector geographic information system (GIS) databases.[6] As a result, public managers "now are able to see the big picture for management purposes, and are better able to understand how actions in one place affect another" (personal interviews, 1999). The partnership has also been behind the initiation of the All Party Monitoring process that is trying to develop an agreed-on set of monitoring protocols and sustainability criteria for the watershed as a whole. This step will assist efforts to measure and monitor progress toward public goals (i.e., strengthen accountability for existing public agencies as well as the partnership; see the section "Monitoring and Measuring: Using Indicators to Promote Transparency" below).

Key examples of how the partnership has managed its dual loyalty to existing mandates and expanded accountability are found in recent agency attempts to manage across the landscape and in the Carberry Creek timber sale—a community-designed alternative that the USFS chose to implement. Each case expands accountability by catalyzing and supporting new ways of doing business that better reflect the unique characteristics of "place" and that recognize more fully the overlapping, integrated character of policy decisions and mandates (across policy sectors as well as political jurisdictions). More specifically, each case expands accountability to the Applegate watershed community by directly involving a broad array of citizens in timber-sale planning and implementation as well as accepting the partnership's challenge to move from a project-by-project management focus to a more holistic "across-the-landscape" focus. Moreover, the empirical record for these two cases supports the idea that the community dimension of accountability can be added *without sacrificing* existing agency commitments to the region and nation.

Seeing the Forest in a New Light: Managing across the Landscape

As part of their holistic ecosystem or watershed management approach, the Applegate Partnership pushes federal agencies to stop managing forests on a project-by-project (timber sale–by–timber sale) basis and to quit focusing so narrowly on only forested lands. Instead, the USFS and BLM are encouraged to manage "across the landscape," incorporating the diverse needs of all flora, fauna, and water resources, as well as the role and impact of human activities on broader watershed health. The belief is that by focusing on and valuing what is being left behind along with what is being harvested (i.e., get the cut out), timber-harvest goals will still be met, while simultaneously addressing the goal of forest health (in terms of biodiversity, stand health, wildlife habitat, and so on).[7] Partnership members also view the new approach as a way to improve the local economy and to reduce the extensive fire hazards that stem in large part from federal agency management practices designed to snuff out all fires, natural or otherwise. On the former point, the partnership wants the agencies to experiment with a diversity of harvest prescriptions, including the harvest of smaller-diameter trees, that benefit local loggers, many of whom have worked only sparingly since Judge Dwyer's spotted owl court decision in 1991.[8] Key to the proposed changes is the insistence that community residents be an active part of the decision process, providing input at every step along the way (Moseley 2000, 15–16; personal interview, 1999).

Gretchen Lloyd, the manager of BLM's Grants Pass Resource Area (RA), was an active participant in the partnership from the very beginning. Under her direction, the Grants Pass RA conducted the first ecosystem restoration timber sale—Ramsey Thin—in 1993. Ramsey Thin was designed to "thin" the forest by focusing primarily on small-diameter trees (those less than 20 inches), thus improving wildlife habitat, reducing fuel loads (hence fire hazards), and promoting more rapid growth of old-growth trees. However, the BLM's Ashland RA, under the direction of area manager Richard Drehobl and planning team leader Steve Armitage, with almost one-third of the land in the Applegate Valley within their jurisdiction, quickly took the lead and demonstrated the possibilities inherent in the new ecosystem restoration approach to timber sales. Data from five timber sales—Lower Thompson Creek, Hinkle Gulch, Middle Thompson Creek, Grubby Sailor, and Sterling Wolf—demonstrate one element of the Partnership's new way of doing business (see table 4.4). Instead of clear-cutting, or marking and cutting only the large-diameter and old-growth trees, the average diameter of trees harvested equaled 12.7 inches, while fully 88 percent of the 223,878 trees were less than 20 inches in diameter. At the same time, although only a few (12 percent) large-diameter trees were cut, they contributed nearly 46 percent of the total timber volume (U.S. Bureau of Land Management, 1999).

The Ashland RA also developed a comprehensive watershed-based planning process that included community members, and experimented with restoration techniques, harvest methods, and timber-sale contracting. The comprehensive planning process selected large areas of ground and viewed them from the ecosystem perspective, "asking what

Table 4.4
Ecosystem-restoration timber sales, 1994–1998, Bureau of Land Management
Medford (Oregon) District, Ashland Resource Area

Diameter in.	Total trees by diameter class	Percentage of total trees	Volume (mbf) by diameter class	Percentage of total volume
08	60,036	26.8	2,705	8.1
10	42,796	19.1	2,687	8.0
12	39,538	17.7	3,498	10.4
14	22,867	10.2	2,837	8.5
16	21,389	9.5	3,683	11.0
18	11,352	5.1	2,701	8.1
20	9,108	4.1	2,984	8.9
22	5,410	2.4	2,430	7.3
24	4,600	2.1	2,823	8.4
26	2,658	1.2	2,212	6.6
28	2,314	1.0	2,670	8.0
29+	1,810	0.8	2,242	6.7
Totals	223,878	100	33,472	100

1. Summary of timber sales in cooperation with Applegate Partnership.
2. The following five sales are included: Lower Thompson Creek, Hinkle Gulch, Middle Thompson Creek, Grubby Sailor, and Sterling Wolf.
3. Average diameter (dbh) of marked trees equals 12.7 inches.
4. There were 6,689 trees marked per million board feet (mbf).
5. This table shows only Douglas fir volume. The total volume of all species for the timber projects was 35,043 mbf.

the land needed rather than what could be extracted from it, though with the understanding that commercial harvest" was still a critical goal (Moseley 2000, 15). The new process was first applied to the Lower Thompson Creek project—a timber sale in which the original BLM plan prescribed clear-cutting for over 90 percent of the acreage (personal interview, 1999). The project focused on both the Douglas fir stands of timber, and traditionally noncommercial small-diameter trees and brush fields. The primary goals were reducing fire hazards, increasing and improving wildlife habitat, and producing a viable commercial timber harvest. The planning team consulted with the general public, conducting several field trips with Partnership participants, interest group representatives, and other local residents, and even going door to door in the immediate neighborhood to facilitate citizen feedback. After doing a "test cut" that followed the management treatment proposed for the larger Lower Thompson Creek timber sale, BLM conducted another series of field trips for Partnership participants as well as interested agency, industry, and environmentalist personnel. After the Boise Cascade timber company started work on Lower Thompson Creek, BLM also agreed to post regular updates of project progress on a billboard at the Applegate Store.[9] Moreover, the updates were sensitive to local residents' quality of life; they alerted residents about more mundane topics such as the scheduling of helicopter flights (noise), logging trucks, and smoke (Moseley 2000, 16).

Given the ecosystem-wide focus and the limited amount of commercial timber in the Lower Thompson Creek project, a chief concern of the Ashland RA planning team was selling the project. In other words, once designed, would anyone purchase the sale and implement the prescribed management treatments?[10] The planning team employed three innovations to make the sale economically viable. First, small-diameter trees were always coupled with enough large-diameter, commercially valuable timber to increase the economic viability of the sale.

Second, instead of commanding the successful bidder to employ a particular logging system—helicopter, cable, tractor, and so on—Armitage and his staff only specified the maximum allowable environmental impact. Any logging system was allowed as long as it complied with the broader environmental requirements of the job. The added flexibility made it easier for loggers to design the most cost-effective extraction system, thus helping to offset some of the higher costs associated with the small-diameter and brush field management treatments. In this case, Boise Cascade selected helicopter logging (personal interview, 1999).

Third, the partnership helped the Ashland RA tap what were called *federal-level 5900 funds* to help finance the sale planning, preparation (e.g., tree marking), and final review. Traditionally, the rules governing 5900 funds defined timber-related restoration so narrowly that they prohibited their expenditure on the "across-the-landscape, ecosystem treatment" desired by the partnership and the Ashland RA. Partnership participants helped the Ashland BLM office successfully lobby the Oregon and national BLM offices for greater flexibility in the application of 5900 funds. From the perspective of an independent scientist and long-time Applegate resident, including the ecosystem-restoration idea as a

Table 4.5
Linking Outcomes to Accountability: The Applegate Partnership

Outcomes	Individual (micro)	Community (meso)	State/ region (mid-macro)	Nation (macro)
Seeing the Forest in a New Light: Managing Across the Landscape	• Strengthens/ clear support for specific livelihoods and/or particular segments of the community (loggers)	• Preserves/ no conflict with laws • Strengthens/ supports positive sum outcome • Strengthens/ diverse array of participants • Strengthens/ increases sensitivity to unique characteristics of "place"	• Preserves/ no conflict with laws	• Preserves/ clear support for laws (NFMA; NEPA) • Strengthens/ extends program reach through innovations to make landscape treatments economically viable • Representatives involved

Note: NFMA is the National Forest Management Act of 1976. NEPA is the National Environmental Protection Act of 1970.

possible candidate for 5900 funds "legitimated the approach as a stand-alone activity in its own right" (personal interview, 1999; Moseley 2000, 18–19).

Lending a Helping Hand: Carberry Creek

The Applegate Ranger District (RD) of the Rogue River National Forest (USFS) has been heavily involved with the partnership's efforts. It provides the Applegate River Watershed Council, a key subcommittee of the Partnership, with rent-free office space and allows the partnership to make copies of meeting minutes on its copier. And the district "has participated in some of the most experimental projects [with the Partnership]"; it is "an incubator, in which interested people [can] try out any number of experiments, [including] . . . small diameter material [timber] utilization and low impact timber harvest methods" (Moseley 2000, 21, 29–30). One of the most important experiments is the Carberry Creek timber sale, an innovative, community-designed, ecosystem management–based forest treatment alternative that preserves and/or strengthens account-ability to individuals, the community, state, and nation (see table 4.6).

The Carberry Creek project grew out of earlier attempts in 1996 and 1997—the Whittenbury Proposal—to write an environmental impact statement for ecosystem management for all federal lands in the Apple-gate.[11] After the Whittenbury Proposal ran into strong resistance by stakeholders, including "territorial" agency types and "environmentalists fearful of the timber harvest volumes the project implied" (Moseley 2000, 26), the Applegate Partnership and the Applegate RD agreed to do a pilot project to keep the spirit of the original proposal alive. The Carberry Creek drainage, selected only after an outreach effort by the partnership determined there was significant interest among residents, covered 7,200 acres. The project itself focused on fire-hazard reduction and on the devel-opment of a detailed landscape-level plan that considers a broad set of goals related to environmental health.[12]

One of the key innovations stemmed from a push by Brett KenCairn from the RIEE, a community organization active in the Partnership, to have people outside the public agencies lead the project. The USFS agreed, in part because the structure of their budget did not allow fund-ing for in-house landscape-level planning, but also because the Apple-gate RD rangers believed they had the discretion to fund environmental assessments conducted by formally established, state-sponsored water-shed groups like the ARWC.[13] KenCairn then acquired additional funds for the landscape planning portion and became project leader, with the Applegate RD playing a critical supporting role through the provision of valuable information (species and archeological surveys) and technical expertise. The Carberry Creek sale was officially contracted for develop-ment in 1998 and completed in 1999.

Water, Gravel, and Salmon

In 1998, the Rural Planning Commission of Jackson County, a county in the Applegate, initiated a periodic review of natural resources using state guidelines provided by the state-level Land Conservation and De-velopment Commission (LCDC). Jackson County's efforts, however, focused only on the aggregate (gravel) resource and were concerned with updating criteria used for identifying and developing all potential

Table 4.6
Linking outcomes to accountability: Applegate Partnership

Outcomes	Individual (micro)	Community (meso)	State/region (mid-macro)	Nation (macro)
Lending a Helping Hand: Carberry Creek	• Strengthens/ clear support for specific livelihoods (loggers)	• Preserves/ no conflict with laws • Strengthens/ supports positive-sum outcome • Strengthens/ diverse array of participants • Strengthens/ increases sensitivity to unique characteristics of "place"	• Preserves/ no conflict with laws	• Preserves/ clear support for laws (NEPA) • Strengthens/ extends program reach by tapping external resources • Representatives involved
Water, Gravel, and Salmon	• Strengthens/ clear support for specific livelihoods (aggregate industry) • Strengthens/ approach increases sensitivity to individual situations (downstream landowners)	• Strengthens/ new data increases general capacity of local government units • Strengthens/ supports positive-sum outcome • Strengthens/ diverse array of participants • Strengthens/ increases sensitivity to unique characteristics of "place"	• Strengthens/ refines program application through new information, improved accuracy (LCDC) • Representatives involved	• Strengthens/ clear support for laws (ESA; CWA) • Representatives involved

Note: NEPA = National Environmental Protection Act; LCDC = Land Conservation and Development Commission; ESA = Endangered Species Act; CWA = Clean Water Act.

gravel-mining sites in the watershed. The Applegate Partnership, along with the Applegate River Watershed Council, became concerned and involved for several reasons. The Rural Planning Commission was not soliciting public input. As a result, many landowners adjacent to gravel pits were not even aware their property rights were on the verge of being restricted. In the words of one citizen, an independent scientist, "they had no idea this process was even going on" (personal interview, 1999). Further, the decision process was treating gravel in isolation, rather than recognizing its relationship and interaction (interconnection) with other natural resources: "They completely denied parity for other natural resources . . . mainly because they had not done any analysis on water quality, riparian areas, and so on. Without this information, they couldn't compare costs and benefits and so their proposal was very single-minded. . . . [It] was always in favor of the aggregate [mining] side" (personal interview, 1999).

Finally, coho salmon, a fish found throughout the Applegate watershed, had recently been proposed for listing under the federal ESA and the commission's proposals were most likely going to hamper coho recovery. For example, most of the gravel-mining sites, or strips, were close enough to streams that they encountered a high risk of flooding during spring runoff, thus contributing significant sediment flows harmful to coho spawning activities and, in some cases, "wildly shifting channel morphology" (personal interview, 1999). Federal officials were also aware of the concern on the part of the aggregate industry, particularly Copeland Sand and Gravel, the primary aggregate contractor in the area, that going ahead with the commission's proposals would be "a nightmare . . . [Copeland] did not want to proceed given the likelihood of such strong opposition from so many people" (personal interview, 1999).

Acting on their concerns, the partnership and the ARWC "found it very easy to mobilize people" and persuaded the Rural Planning Commission (RPC) to listen to citizens' concerns. The ARWC, with vocal support from partnership participants and others in the audience, offered "to help the Commission get the information on the other resources so that they could see the bigger picture and develop a more balanced approach to dealing with the aggregate issue" (personal interview, 1999). After the planning commission agreed it was worth a try, the ARWC convened a public meeting to discuss the possibility of collecting the data and managing the aggregate issue in such a way that virtually everyone involved came away happy. More than 100 people attended the meeting and many signed on as volunteers for the length of the project.[14] The working group included landowners representing their property-rights interests, the aggregate industry, the county planning department, the Watershed Council, agricultural interests, environmentalists, and Applegate Partnership participants. In the end, the effort resulted in a comprehensive natural resource inventory and a review of, and suggested reforms for, all county-level ordinances related to natural resources. The reforms were designed to promote a more holistic, balanced approach to natural resource decision making and a more balanced consideration of the rights of landowners owning non-aggregate-bearing property adjacent to the mining sites.

Accountability occurs at all four levels in the "water, salmon, and gravel" case (see table 4.5). The case addresses the concerns of individual property owners, especially nonaggregate and downstream properties. The cooperative agreement meets the desire of the aggregate industry to avoid constant battling (transaction costs) and, according to a federal official, has helped it "to improve its neighborly image" (goodwill) (personal interview, 1999). The county gains improved information databases for managing natural resources. Rural citizens benefit because the decision process engaged hundreds of rural citizens rather than only an elite few. The effort also provides accountability to the state by adhering to the spirit of the regulations managed by the LCDC, although by taking a comprehensive approach to natural resources rather than a fragmented one. Moreover, Applegate residents proactively put forth a solution that attempts to achieve the results implied by the federal ESA and the Clean Water Act (CWA) by incorporating information on water quality, riparian ecosystem needs, and other items with direct impacts on the health and survivability of steelhead and salmon.

A Little Bit of Honey Goes a Long Way: Helping Neighbors Become Better Environmental Stewards

A key premise of the Applegate Partnership is that many (not all) people, especially farmers, ranchers, and loggers with small operations, may only appear to be antienvironmental (i.e., they exhibit a consistent pattern of behavior that contributes to environmental degradation). According to this line of reasoning, closer inspection is likely to reveal at least some people who are unaware of the extent of the damage their actions cause, lack knowledge regarding potential cleanup solutions, and/or lack the resources to implement solutions to minimize or eliminate the environmental damage associated with their operations. In such cases, honey—persuasion, information, and resource assistance—rather than vinegar—coercion, labeling the violator as an unwanted member of the community—might do a better job of getting win-win-win results consistent with the "environment, economy, and community" mission. As one environmentalist put it: "Farmers and loggers are a vital economic component of our community, but many of the smaller operations are also marginal, economically speaking. . . . Simply because they pollute should we call out the dogs [EPA], put them out of business, and throw them out of the community? No, we should build bridges to them and help them become better environmental stewards. . . . After all, they're neighbors, too" (personal interview, 1999). In another case, new information derived from participation in the partnership helped a farmer recognize the intimate dependency between bats—a form of "night-flying insect control" essential for maintaining crop health—and "old tree snags that we always felt were expendable. We found out that some of them house [the very] bat communities [that help us]. Now I see snags in a different light" (personal interview, 1999).

A classic case where this philosophy was put into action involves a stockyard business in the middle of the Applegate watershed. The owner winters cattle, yearlings, and other dairy cows for a number of farmers in the region in stockyards that, according to an environmentalist, "looked like a nuclear waste site . . . a lot of cows, a lot of muck and mud" (personal interview, 1999). The site was both an eyesore and an environmental problem because the "muck and mud" migrated into the stream running through his property, creating high sediment loads and water quality problems detrimental to aquatic species, including fish. Many environmentalists and anglers who liked to fish the stream below the property wanted to solve the problem by calling in the EPA and the state Department of Agriculture to severely restrict or close down the stockyard.

Partnership board members were not convinced this would solve the problem, in large part because the cattle, and the environmental problem, would simply shift to another location. Moreover, it would remove a viable economic component of the community. The partnership decided to "talk to [the owner] and ask what's going on. . . . We found out he was concerned about the ecological issues, but he didn't have the money to do anything about it and he didn't know what to do about it anyway. . . . We . . . ended up finding a way to fund some of the restoration work to the tune of $17,000. We put up berms on the sides of the creek [to keep the cattle away], built about a mile of tree-planted riparian areas, and planted some hard rock on the bottom of the streambed where the cattle crossed to minimize silt [load] problems. . . . [A group of] fly-fishermen [actually] came in and planted the trees" (personal interview, 1999).

In the end, a citizen-member of the Partnership notes that the effort "did not solve as much of the environmental problem as [we] had hoped; . . . it was a Band-Aid solution" (personal interview, 1999). But it was a step in the right direction. The stockyard owner changed some of his management practices by restructuring the physical environment of the stockyard in relation to the stream. The end result was a healthier riparian habitat than before. Just as importantly, it "built a bridge to the farming community" and "increased the potential for more agricultural interests to adopt more environmentally sensitive management practices" (personal interviews, 1999). More generally, a farmer-participant observes that the partnership's "willingness to work together as neighbors and to accept farming as a legitimate enterprise" is attracting leaders within the agricultural community to the Partnership, including Connie Young, chair of the Josephine County Soil/Water Conservation District, a board member of the Josephine County Farm Bureau, and an Applegate Partnership board member representing agricultural interests (personal interview, 1999). The partnership "is [also] excited that Offenbacher, [the scion of a family that has farmed in the Applegate for over 100 years,] is joining the board" (as of summer 1999) (personal interviews, 1999).

In the "neighbors-helping-neighbors" case there is accountability to the individual needs of particular community members, accountability to community because it is a cleaner place to live and because the fisheries resource is likely to improve, and accountability to state and federal environmental laws (see table 4.7).[15]

Table 4.7
Linking outcomes to accountability: Applegate Partnership

Outcomes	Individual (micro)	Community (meso)	State/region (mid-macro)	Nation (macro)
Helping Neighbors Become Better Environmental Stewards	• Strengthens/clear support for an individual's problem (financial subsidy, catalyzes other private-sector, labor-based resources for a private landowner)	• Preserves/no conflict with laws • Strengthens/supports positive-sum outcome	• Preserves/clear support for laws (environmental protection)	• Preserves/clear support for laws (environmental protection)
Endangered Species and Connecting the Private-Public Lands Mosaic Together	• Strengthens/approach increases sensitivity to individual situations (private landowners)	Preserves/no conflict with laws • Strengthens/supports positive-sum outcome • Strengthens/diverse array of participants	Preserves/clear support for environmental protection goals • Strengthens/extends program reach to private land • Representatives involved	Preserves/ clear support for laws (ESA) • Strengthens/extends program reach to private land • Representatives involved
Local Exchanges and New Markets	• Strengthens/clear support for specific livelihoods and/or particular segments of the community (farmers, organic and traditional; individual consumers pay less and enjoy more choices)	• Preserves/clear support for community goals • Strengthens/supports positive-sum outcome	• Preserves/clear support for state goals (economic development)	• Preserves/no conflict with laws

Connecting the Private-Public Lands Mosaic Together for the Sake of, among Other Things, Endangered Species

At first glance, the overwhelmingly public character of the land-tenure pattern in the Applegate watershed—70 percent is owned by either the USFS or the BLM—suggests that integrating private lands into the natural resource management mix is of relatively minor concern to watershed and species health.[16] From this perspective, logic dictates that most of the management effort should concentrate on public lands. A federal official involved in the Applegate Partnership argues, however, that such an approach "is deceivingly simplistic" (personal interviews, 1999 and 2000). Significant portions of the federal public lands, particularly that owned by the BLM, exist only as fragmented pockets of land strung together in checkerboard fashion, thus complicating *publicly* led efforts to manage for biodiversity, water quality, and other elements of ecological health (personal interview, 1999). In addition, an environmentalist notes that "the majority of the prime habitat for the majority of the plant and animal species here in the Applegate is on private land because it occupies the lowlands and riparian zones. So, . . . let's just say that the BLM and Forest Service manage their lands ecologically perfectly, whatever that is. My question would be: So what? . . . We're still not going to save the vast majority of these critters without the cooperation of the private landowners. . . . They have got to be involved" (personal interview, 1999). Moreover, a federal official, as well as other partnership participants, argue that the federal ESA, designed as it is to prevent losses of plant and animal species, and to preserve habitat essential to endangered animals,[17] "does nothing to *improve* habitat, to better promote the ability of endangered species to flourish and grow. . . . [The problem is] that a 'stop the losses' approach can stop people from building on or destroying habitat, . . . [yet] potential [ESA] listings [can] also promote a shoot, shovel, and shut up response by private landowners" prior to the legal listing, thus hastening the decline of endangered species (personal interviews, 1999; Mann and Plummer 1995).

Many officials in the federal land agencies are in agreement with the partnership's perspective; they recognize the need to integrate private and public land management in order to secure an effective management regime for natural resources.[18] But in the Applegate, as in many other places of the rural West in which the federal government plays a dominant role, public land managers run head long into a typical Western sentiment—government stay out and off of my land.[19] Given that "distrust of the federal government is rampant," agency officials are not in a good position to coordinate such management efforts, much less open the lines of communication necessary to getting private lands into the management mix (personal interviews, 1999). This is where the partnership steps in. It provides a new venue for communicating with private landowners and persuading them of the need for coordinated, integrated management and the importance of managing private lands so as to benefit the environment. For example, the deliberative and education components of the partnership can not only transform people's perspectives (see section on "transformation" below), they can also create "new trust" among traditional adversaries and a greater willingness on the part of private parties to cooperate with publicly defined goals (personal interviews, 1999). (See table 4.6.)[20]

Public agencies also "lean on the partnership" by recognizing and relying on certain community assets, specifically the established personal relationships and influence partnership participants have with other private landowners: "We are kind of a go-between with the agencies and the landowners . . . [because] it's easier for us to get onto private property because they're our neighbors and, in some cases, we've known them for ten, twenty, thirty years or more" (personal interviews, 1999).

At the same time, Shannon, Sturtevant, and Trask (1997, 10) note that "agency personnel in the Applegate find that their interaction with the community through concrete examples and demonstrations (e.g., field trips and booths at community days) and communal activities oriented toward improving the watershed (e.g., tree planting and nursery plot thinning) not only improve communication and learning, they build social capital and social solidarity—important bases for collaboration and trust."

For most participants, however, the key to connecting the private-public lands mosaic together is the state-sponsored but locally managed Applegate River Watershed Council (ARWC). Originally conceived prior to state legislation as a subcommittee of the partnership in 1992, the watershed council is viewed by many as "the unofficial implementation arm of the partnership" (personal interviews, 1999). Even today, it stills retains its status as a subcommittee of the partnership.

The ARWC is an advisory body that seeks to improve water quality and the overall management of water resources in the watershed. The Council has sponsored a number of watershed health projects, including, among other things, new irrigation headgates to better manage water flows, in-stream habitat enhancement, road restoration, and tree planting in riparian zones. They have completed watershed assessments of the entire 500,000-acre Applegate River watershed (in 1994) and a more detailed look at the Little Applegate River watershed, a 72,000-acre subbasin in the larger Applegate River watershed (in 1998). The ARWC has also worked with state and national agencies as well as private-sector parties to scientifically evaluate historical flooding and hydrologic conditions along the lower Applegate River floodplain (a river-systems analysis). In the Cheney Creek subwatershed, the Council has worked with BLM and the Oregon Department of Forestry to evaluate watershed conditions across private and public lands and to promote stewardship and ecological awareness. Another example involves improving fish passage. The watershed offers prime spawning habitat to steelhead and coho and chinook salmon—all migratory fish. But concrete and gravel "push-up" dams,[21] poorly placed culverts,[22] and irrigation ditches that capture stream flows[23] can prevent such fish from migrating upstream to spawn or downstream to quality freshwater or ocean rearing habitats. The ARWC is identifying and cataloging the in-stream barriers to fish passage and crafting solutions for improving them (Applegate River Watershed Council, 1998).

A participant from a federal agency argues that "because [the ARWC] is much better connected to the community itself, it is able to get private lands into the management mix without pissing everyone off. . . . It's [also] well-positioned to manage collective, watershed-wide problems" (personal interview, 1999). In large part this is because it is the only entity with a management focus covering the entirety of the watershed *and* that, according to an environmentalist, "practices proactive ecological enhancement, which tries to restore and improve habitat, rather than simply stopping [environmental] degradation and the destruction of habitat" (personal interview, 1999). Currently, there are no incorporated towns in the Applegate, three counties with political jurisdiction, and three major federal management units (two BLM Resource Areas and the Rogue River National Forest). Moreover, despite the fact that most of the Applegate is included in the federal Applegate AMA under the Northwest Forest Plan and that many agencies (state and federal) "have been working really hard on looking beyond their own agency," federal attempts to manage the land inclusively are still plagued by traditional fragmentation and a general lack of will to pull off such an integrated approach. State agencies are experiencing the same trouble, according to a citizen-scientist: "We were shocked when we first sat down with the state management agencies. [Individual agencies] didn't exactly understand what the other state agencies were doing. . . . They didn't have a clue. . . . Nor are they practiced at working together" (personal interview, 1999).

According to a timber industry official who has participated in the Applegate Partnership, this is where the ARWC shares a key characteristic with the partnership; the ARWC "adds value by being a place where bridges [between agencies] can be built and coordination can occur" (personal interview, 1999). Together with the partnership, the watershed council performs a role as a catalytic, coordinative core that helps to facilitate efficient governance by decreasing program redundancies and providing a one-stop communication forum for the many stakeholders in the area, including government agencies. In this regard, the partnership and the ARWC display little interest in being able to claim sole responsibility for making the watershed cleaner or for preserving critical habitat, than in making sure that previously defined local, state, and federal goals are achieved. In short, the two do not have to actually "do" every project themselves; rather, they often seek to coordinate and catalyze others' resources in the name of more effective watershed management because "no one group or agency can do it alone" (personal interview, 1999).

Promoting Local Exchanges and New Markets

The partnership promotes the "economy" part of its mission by helping to create exchange networks and contact lists that connect people and businesses within the watershed to each other (as opposed to an economic development strategy relying on financing). The primary focus is on creating new producer-to-producer and producer-to-consumer relationships within cottage industries, even to the extent of encouraging new businesses to form. For example, there is the Applegate Core, a spin-off from the partnership. It started out as cottage industries getting together about once a month for potluck dinners in order to learn what kinds of products others are selling and to see what kinds of direct (no intermediary) producer-to-consumer transactions are viable. The direct relationships enable farmers, for example, to sell their products for more, while saving consumers money at the same time. Such a direct relationship also gives purchasers the satisfaction of knowing exactly where their food comes from and how it is grown (i.e., what kinds of pesticides, fertilizers, hormones, and so on were used). Some ranchers and farmers now sell "farm fresh beef right off the farm. . . . We don't use any additives or beef product, it's an animal that has been fed good wholesome hay and grass. Consumers see the animal, they purchase the animal at hanging weight, then we have someone we really trust, Leon from Medford, come in and slaughter the animal right here in our barnyard. Then we take it to Cartwright Meatpacking and the person who bought the animal pays for the cut and wrapping. And they tell them how they want it cut up and so on and so forth" (personal interview, 1999).

Another spin-off from the partnership, the Applegate Aggregate, focuses on linking certified organic farmers together to share best practices and to market produce directly to consumers in the Applegate as well as in the adjacent areas of Medford and Grants Pass, Oregon. Further, the partnership, like the BLM and USFS, now encourage the development of specialty wood products from the timber left over from traditional logging operations as well as sales that "thin" forests from below (clear out the understory to promote more rapid growth of remaining trees). Surplus timber includes small-diameter trees, noncommercial-grade wood, and brush remnants—items that are not salable in the traditional commercial timber market. Applegate residents are turning the surplus into birdhouses, animal toys, roughhewn furniture, wall hangings, bird perches, and other assorted arts and crafts products and shipping them all over the United States (personal interviews, 1999).

The promotion of local exchanges and new markets primarily focuses accountability at three levels—the individual, community, and state (see table 4.6). The efforts tap existing community assets to maintain and improve the local *rural* economy, a key concern for local and state law-makers. As a result, job opportunities are enhanced, consumers save money, and producers make more. The added profits for agricultural producers help the owners of small family farms, many of which are several generations old, combat the combination of low commodity prices and rising real estate values and taxes that are leading more people to subdivide and develop their land for the upscale home market. Farmers thus enjoy additional opportunities to maintain their economic viability and meet a key goal of many farm families, according a farmer, and that is to "keep the farm in the family for another generation" (personal interview, 1999). At the same time, residents in and around the Applegate who have a preference for farm-fresh and/or organic foods encounter a wider variety of choice and greater ease in pursuing their preferred lifestyle.

Further, the types of economic development promoted by the partnership and affiliated efforts work to preserve the rural character of the Applegate—a key goal for community members as expressed in opinion poll surveys (Priester 1994) and a more formalized community assessment process (Priester and Moseley 1997).[24] Promoting existing agriculturally based land-use patterns and small cottage industries rather than attracting new, larger production facilities and capital from outside the area is only part of the story, however. The other side of the same coin is a desire to slow the rate at which rural land is subdivided for 5,000- to 10,000-square-foot trophy vacation and "second" homes. Participants, including a citizen unaffiliated with any organized group, are concerned that such homes not only "require a small forest worth of timber to build," they are typically occupied for only a few weeks each year by absentee owners (personal interviews, 1999; Priester and Moseley 1997).

Monitoring and Measuring: Using Indicators to Promote Transparency

The Applegate Partnership has always been sensitive about making sure that their proceedings and decisions are open and accessible to community residents as well as to any interested parties from outside the area (personal interview, 1999). Developing the capacity to monitor and measure results is viewed as a critical component of this larger effort (personal interviews, 1999).

For all these reasons, the Partnership has established an information, research, and monitoring committee and conducted hundreds of field trips to inspect projects before, during, and after they are implemented. They also have commissioned a number of communitywide studies designed to increase understanding of the area's social, demographic, economic, and environmental characteristics and to canvass residents for their views on the most important issues and problems affecting the Applegate. Chief among these studies are Priester's (1994) community assessment, an economic and demographic assessment (Reid, Young, and Russell 1996), and a strategic planning document (Priester and Moseley 1997). The USFS and BLM have likewise contributed significantly in this area, publishing an ecosystem health assessment (U.S. Bureau of Land Management and U.S. Forest Service, 1994), a description of what ecosystem restoration looks like (U.S. Bureau of Land Management, 1998), and the Applegate Adaptive Management Area guide (U.S. Bureau of Land Management and U.S. Forest Service, 1998).

But problems still exist because there is conflict among scientists, agencies, private-sector interests, and environmentalists over appropriate monitoring protocols and criteria (indicators) for measuring sustainability as well as agency-mandated goals. An agency participant argues that "sustainability is a critical goal for the partnership . . . . But how are we going to get there? What's really important about it? What is it we're trying to reach for? . . . If we can't translate that into some criteria, agree on what [the criteria] are, and then have the ability to monitor it so we can learn for the future, I think things are going to just keep hitting chuckholes" (personal interview, 1999; Applegate River Watershed Council, 1998).

The problem is not with the amount of environmental monitoring going on; "there's an incredible amount of that being done by the ARWC, USFS, BLM, you name it" (personal interview, 1999). The problem is that historically everyone collects and manages their own databases on a separate basis: "The Forest Service has their information, the BLM has their information, NMFS has its information, and so on. . . . Everybody has their own bits of information, based on their perspective of the truth—whatever it is. What [the partnership] would like to do is develop a mechanism by which every piece of monitoring information about the

Applegate is done in an integrated fashion. Whether it's done by the Forest Service, BLM, EPA or whoever, but it all goes into a central, integrated database" (personal interview, 1999).

In an attempt to overcome these problems and to increase the transparency of their efforts, the partnership held a series of meetings focused on what is known as all party monitoring, or APM, in the spring of 1999. The effort seeks to bring together all the parties with a stake in the Applegate watershed to develop agreed-on monitoring protocols and sustainability indicators suited to the unique needs of the watershed, yet still consistent with the mandates of the national agencies that exercise jurisdiction over most of the area (Applegate River Watershed Council, 1998).

The monitoring efforts, especially when coupled with the allegiance to inclusivity and the open-access design, signal that governance transparency is a priority for the partnership. It also signals that accountability to all four levels—individual, community, state/region, and nation—is central to governance efforts (see table 4.8).

Transforming Individuals' Worldviews

The Applegate Partnership is in the process of building institutions that *may* govern, or have substantial effect on, citizens' behavior, decisions, and outlook toward others and their community. To what extent, for example, do the new institutions create new relationships, foster a greater degree of trust among citizens, and cultivate a heightened sense of collective purpose in the watershed that is centered on the tripartite "environment, economy, and community" mission? Put differently, are individuals' worldviews being transformed with respect to how they view their neighbors, their preferences for policy, and their role in natural resource management? And how are such transformations, if any, connected to the accountability equation? Are they effecting an increase or decrease in commitment to broad-based accountability (i.e., accountability to community, state/region, and nation? (See table 4.7.)

The partnership focus on deliberation and cooperation with others on projects providing collective benefits has created new, constructive, trust-based working relationships. Participants were asked to assess whether

Table 4.8
Linking outcomes to accountability: Applegate Partnership

Outcomes	Individual (micro)	Community (meso)	State/region (mid-macro)	Nation (macro)
Monitoring and Measuring: Indicators and Transparency	• Weakens/focus is on collective health of community • Weakens/increased transparency makes it harder to promote zero-sum outcomes	• Strengthens/new data increases general capacity of local government units • Strengthens/supports positive-sum outcome • Strengthens/diverse array of participants • Strengthens/increases transparency and ability to monitor government/governance decisions	Preserves/no conflict with laws • Strengthens/increases transparency and ability to monitor government/governance decisions	• Preserves/no conflict with laws • Strengthens/increases transparency and ability to monitor government/governance decisions
Changing Worldviews	• Weakens (people shift toward the idea of enlightened self-interest and a focus on win-win-win outcomes)	• Strengthens	• Strengthens	• Potentially strengthens, but unclear

their participation in partnership proceedings had changed their willingness to trust others, especially those they may have viewed as adversaries prior to joining the partnership. In every instance (twenty interviews) the answer was yes, participants now trust others, even adversaries, more than before. In the eyes of a federal official, not only "has there been growth in the relationships, there has been growth in understanding of each other's issues, and more respect for each other's issues and positions" (personal interview, 1999). An environmentalist adds that "rather than being adversaries we've come to realize that we have a great deal in common. We still disagree on lots of things, but the partnership helps us understand that we're also neighbors with a common stake in the health and well-being of our community" (personal interview, 1999).

At the same time, a number of citizen-participants pointed out that their trust in government has increased, a fact noticed by agency officials (personal interviews, 1999). Government officials found that their trust in the public increased, in large part because they found that community residents were far more interested in achieving a collective good as opposed to "being out for themselves or for their little group" (personal interview, 1999). One official explains that

the partnership laid the groundwork for trust. It gave me time and experience with others in the community, . . . the very act of sitting down that much over the years increases trust [because it can and did build] . . . deep friendships. I came to trust and place great value in what community people had to offer. And I came to trust that our agency projects would benefit from more interaction [with the public]. (personal interview, 1999)

Most participants are careful, however, to *individualize* their trust of others. There is a clear sense that the openness and iterative deliberations of the partnership make it easier to discern who is worth trusting and who is not, with a premium placed on forthrightness, integrity, and honesty. For example, one environmentalist finds it much easier to trust individuals in the government, but "as institutions, the BLM and Forest Service are not quite there yet" (personal interview, 1999). Another longtime participant from the farming community says, "I've learned to trust many of the people in the partnership because they're right upfront with their ideas and their philosophy. . . . I know that if they give their word on something, it's true. They're not hiding anything. . . . If you're going to be in the partnership, you have to be upfront and honest with your feelings and your philosophy. And the rest will respect you for that" (personal interview, 1999). In another case, a government official explains that "it all ties back to individuals. Through the partnership effort I've come to know people better and have identified people who I feel like I can work with and be more trusting of. But the reverse is also true" (personal interview, 1999).

Taken by itself, the increase in trust among community members helps to facilitate governance effectiveness (Fukuyama 1995; Putnam 1993), but it does not necessarily translate into greater accountability to collective goals beyond the community level. Yet the willingness to trust appears essential to the kinds of individual transformations expected by democratic theorists (e.g., Dawes, van de Kragt, and Orbell 1990; Kemmis 1990; Piore 1995). And although there is some debate among democratic theorists over whether deliberative, participative venues can actually transform people into thinking and behaving more broadly, or collectively (Warren 1992; Morrell 1999), it is interesting that many of the participants themselves are "absolutely" convinced that their participation in deliberative, consensus-based governance arrangements has changed their outlook and made them more sensitive to collective concerns. Many who have been involved with the partnership for years, in fact, do not understand why this is such a mystery to academics (personal interviews, 1999 and 2000).

Evidence from twenty interviews suggests that a significant number of participants are now more willing to think of their own individual/personal situations as connected to, or as an extension of, the larger whole (rather than viewing issues and preferred outcomes from a narrower, more self-interested perspective). When asked whether participation in the Applegate Partnership has led them to give greater weight to how proposed actions will affect *the world outside of the watershed community,* fully 40 percent of those interviewed said yes (eight out of twenty). Interestingly, 50 percent of those interviewed claimed the willingness to consider the effect of proposed partnership decisions on the outside world *as a starting point* (i.e., the institutional dynamic matched or reinforced their original position). When asked whether participation has led them to give greater weight to the benefits of proposed actions for the *watershed community,* 45 percent answered yes (nine out of twenty), while 30 percent claimed community-mindedness as an original position.

Conclusion

The big picture that emerges from the outcomes produced by the Applegate Partnership is one that complements and reinforces the broad-based, simultaneous accountability dynamic produced by its institutional structure, processes, norms, and management approach. In this respect, too, the partnership is accountable to a broad cross-section of society.

The two strongest cases—"Water, Gravel, and Salmon" and the "Connecting the Public-Private Lands Mosaic"—strengthen accountability across all four levels. Another two, the "Carberry Creek" and "Managing Across the Landscape" cases, simultaneously strengthen accountability to the individual, community, and national levels, while still preserving accountability to the state level (a case of no conflict). The "Monitoring and Measuring" case similarly strengthens accountability to the three collective levels—community, state, and nation—while weakening accountability to individuals. In yet another combination, the "Neighbors Helping Neighbors," "Diversity of Representation," and "New Exchanges and Markets" cases successfully strengthen accountability to the individual and community levels at the same time that accountability is preserved for the state and national levels. The final case—"Changing Worldviews"—weakens accountability to individuals, while concurrently making accountability stronger at the community and state levels.

Nor do the cases taken as a whole support the critics' contention that decentralized, collaborative, and participative governance arrangements lead to industry domination and an increased propensity to roll back state and federal laws, especially environmental laws, already on the books. The lesson gleaned from the meeting-level participation statistics, namely, that environmentalists play an instrumental role in Partnership proceedings, finds strong support in the other cases.

In fact, in all the cases where the Applegate Partnership promotes stronger accountability to individuals, whether it is loggers, the aggregate industry, or private property owners, the accountability does not come at the expense of either collective interests or other levels of accountability. Instead, the increase in individual accountability occurs within the context of comparable accountability gains for the community, the state/region, and the nation, most generally in the form of positive sum outcomes, support for existing laws, or stronger, more effective government programs.

5

Coping with Conflicting Water Resource Demands in the Henry's Fork Watershed

It is hard not to marvel at the towering, jagged precipices comprising the Teton Mountains that border the east side of the Henry's Fork watershed. They appear as an alien, distant land form that is a far cry from the orderliness and apparent tranquility encountered among the almost 2,000 farms populating the watershed's landscape. Yet it was not that long ago that the political and social dynamic governing the major interests and agencies with stakes in the Henry's Fork watershed mimicked the raw jaggedness of the Tetons. Stakeholders were at each other's throats clamoring for policies that necessarily favored some at the expense of others, demanding rather than asking that their preferences be codified into law, fighting incessant court battles, and ostracizing those who dared to differ.

Over the last seven years, however, the river of conflict has slowed to a trickle as stakeholders from across the board have come together for the sake of preserving and improving the Henry's Fork watershed. The stories of conflict and bitter partisanship are starting to be replaced by stories of friendship, mutual respect, innovation, and outcomes that purposely promote the multiple objectives of the various stakeholders in mutually beneficial, positive-sum ways. At the center of this new universe is the Henry's Fork Watershed Council (HFWC), practicing civility, promoting cooperation among neighbors, and, if rhetoric is to be believed, insisting on outcomes that serve a broad public interest. Yet even such an ardent proponent of the watershed council as Jan Brown, a founding member and one of the HFWC's two lead facilitators for the past seven years, willingly admits that "none of this would have happened if Dale Swenson and the Fremont-Madison Irrigation District [FMID] had not agreed to share power with the rest of us. They hold the cards [i.e., senior water rights] and they decided to share power. And because of this, and despite the fact that we used to be virtually mortal enemies, we have now worked together for more than six years and I now count Dale as one of my best friends, a true friend who has integrity, is honest, and cares deeply about the Henry's Fork watershed community" (public remarks, Eastern Idaho Watershed Conference, October 20, 1999).

A skeptic might interpret such a remark as coming from one who has been co-opted and is so wrapped up in making the effort succeed that they can no longer objectively observe, much less understand, the true course of events. The underlying truth, from this perspective, is that decentralized, collaborative, and participative governance arrangements not only are not designed to bring about broad, simultaneous accountability, in actuality they are nothing more than a sophisticated facade for the more significant hidden dynamic. To wit, commodity interests are simply using the HFWC to slowly, inexorably mount an assault on environmental laws and to return control over policy governing natural resources to those with the greatest direct stake in land management: the farmers, irrigators, ranchers, and loggers.

Which of the two accounts best represents the reality of the HFWC? Is community being restored and, along with it, broad-based accountability? Or are the skeptics right? The logic of the skeptics' argument suggests that understanding the connection between accountability and the institutional structure, process, participant norms, and preferred holistic approach to management is simply not enough. Something more is needed.

One way to gain some perspective on this debate is to examine a series of the actual outcomes generated by the HFWC. Eleven outcomes ranging from diversity of representation to helping agencies achieve their goals, dam privatization, and changing worldviews are described and mapped against the individual, community, state/region, and national levels of accountability. Outcomes that meet the demands of a broad-based, simultaneous form of accountability display a win-win-win character. In other words, there is accountability to multiple interests and levels all at the same time. If, on the other hand, the criticisms are valid, we should expect to see a large imbalance in the way that accountability is apportioned among the levels. Most outcomes should end up strengthening accountability to the individual and/or community (meso) levels at the expense of the state/region and national levels.

Diversity of Representation

The roster of participants in the HFWC ranges across virtually the entire spectrum of interests in the watershed region, including state/regional and national interests. Integral participants come from a broad variety of environmental organizations, state- and federal-level agencies concerned with the management of public lands and natural resources, and local administrative officials such as planning and zoning or vegetation/weed control personnel. There also are independent and university-based scientists, ranching interests, recreation interests focused on fishing, hunting, and off-road motor vehicle use, farmers and irrigators, and visitors from outside the watershed area. Table 5.1 displays the distribution of participants for all HFWC meetings from January 1994 through March 2000. A total of 325 people have attended at least one of the 46 meetings, with an average of 44 people in attendance at each meeting.[1] Despite the small number of participants in relation to overall community size (roughly 1 percent of the area's total population), there is balanced participation among a broad cross-section of interests (table 5.1) and a remarkable balance in representation among the different levels of accountability (see table 5.2). It thus is plausible to argue that the HFWC case enhances accountability to a broad cross-section of the community, while simultaneously preserving a strong measure of accountability to the state/regional and national levels. Accountability to individuals also is strengthened given the repeated opportunities to engage the policy process, although it is clear that only a relatively small number of citizens take advantage of these opportunities on a regular basis (see table 5.3).

The most basic level of analysis in Table 5.1 considers all 46 meetings. For all HFWC meetings, there are significant levels of participation among four key groups with major stakes in the watershed. Federal administrative agencies and state-level natural resource bureaucracies—organizations that are manifestations of national and state mandates—have average participation rates of 19.7 percent and 13 percent, respectively. There is also a relative balance in participation by commodity interests as opposed to environmentalists. Commodity and business interests comprise almost 17 percent of those in attendance, while environmentalists constitute 13.1 percent of all participants. Among other categories, concerned citizens, or those without any formal group affiliation, lead the

Table 5.1

Distribution of participants in the Henry's Fork Watershed Council, all meetings from January 1994 through March 2000

Participant category	Total attendance (46 meetings) $n = 325$ %	Attends 25% or more of all meetings $n = 60$ %	Attends 50% or more of all meetings $n = 27$ %	Attends 75% or more of all meetings $n = 7$ %
Environmentalists (Henry's Fork Foundation;* Idaho Rivers United; Idaho Wildlife Federation; Greater Yellowstone Coalition; Nature Conservancy; Teton Regional Land Trust)	13.1	15.0	18.5	14.3
Extractive/commodity interests (farmers, irrigators, ranchers, timber, local business and development interests)	16.9	16.7	22.2	42.9
Recreation interests (Henry's Fork Foundation; Idaho Rivers United; Rocky Mountain Elk Foundation; Blue Ribbon Commission)	8.0	5.0	7.4	14.3
Federal administrative officials (Bureau of Land Management; Bureau of Reclamation; Fish and Wildlife; Natural Resources and Conservation Service; Forest Service; U. S. Geological Survey)	19.7	28.3	14.8	0.0
State administrative officials (Division of Environmental Quality, Departments of Fish and Game, Parks and Recreation, Public Lands, and Water Resources)	13.0	20.0	22.2	28.6
Local administrative officials	2.5	3.3	0.0	0.0
Federal legislative representatives	1.6	0.0	0.0	0.0
State legislative representatives	0.3	0.0	0.0	0.0
Local (elected) officials	1.6	0.0	0.0	0.0
Concerned citizens (no formal group affiliation)	10.8	1.7	0.0	0.0
Independent and university-based scientists	7.5	8.3	11.1	0.0
Media	5.0	1.7	0.0	0.0
Total Attendance	100	100	100	100

*The Henry's Fork Foundation and Idaho Rivers United representatives are apportioned equally to two categories—environmentalists and recreation interests—given their interest in fishing as a sport.

Table 5.2
Distribution of attendance by location, all meetings ($n = 325$)

Participant category	Local %	Regional/ state %	National %	Totals %
Environmentalists	5.5	6.7	0.9	13.1
Extractive/commodity interests	12.4	4.1	0.4	16.9
Recreation interests	4.6	3.4	0.0	8.0
Administrative officials/bureaucratic experts	2.5	13.0	19.7	35.2
Legislative representatives (elected or staff)	1.6	0.3	1.6	3.5
Concerned citizens	4.7	5.2	0.9	10.8
Independent and university-based scientists	2.5	4.1	0.9	7.5
Media	1.6	3.4	0.0	5.0
Total Attendance	34.4	39.4	23.9	100

Table 5.3
Linking outcomes to accountability: Henry's Fork Watershed Council

Outcomes	Individual (micro)	Community (meso)	State/region (mid-macro)	Nation (macro)
Diversity of Representation	• Strengthens/ open-access, iterative design equals greater opportunities for individual citizens	• Strong/a broad cross-section of community-based stakeholders	• Preserves (see table 2)	• Preserves (see table 2)
Staying True to Agency Mandates: Grizzly Bears and Roads	• Strengthens/ outcome is zero sum • Strengthens/ clear support for specific livelihoods and/or partic-ular segments of the commu-nity (environ-mentalists)	• Preserves/ no conflict with laws • Weakens/ community not involved in decision	• Preserves/ no conflict with laws • Weakens/ state not involved in decision	• Preserves/ clear support for laws (ESA) • Repre-sentatives involved/ they decide the issue

way with almost 11 percent of all participants, with recreation interests registering a participation rate of 8 percent and scientists/academics (both independent and university-based) at 7.5 percent. In addition, participants from the electronic and print media average 5 percent of all participant visits. There are also occasional visits by local administrative officials (2.5 percent), local elected officials (1.6 percent), and legislative staffers for elected representatives from the U.S. House and Senate (1.6 percent).

Columns 2 through 4 in table 5.1 examine how the frequency or intensity of attendance by individual participants affects the balance of voices at HFWC meetings. Intensity of participation is measured in three categories—those attending 25 percent or more of all meetings ($n = 60$), those attending 50 percent or more of the meetings ($n = 27$), and those exhibiting the highest level of commitment by attending 75 percent or more of all meetings ($n = 7$).

Beyond the immediate observation that the core groups of repeat players are much smaller than the total number of HFWC participants over time, there are only a few major changes in representation patterns in the first category (attends 25 percent or more of all meetings). There is the same relative balance among environmentalist, commodity, and recreation interests, while there is a significant increase in the share of participation by federal and state administrative officials. There are also two significant declines. Participation by unaffiliated citizens drops from 10.8 percent to less than 2 percent, and media participation declines to less than 2 percent. Moreover, the expectation of GREM critics that commodity interests might dominate decentralized, collaborative arrangements does not find any support until the third category (75 percent or more of all meetings), where three of the seven participants are from the commodity category. Yet given the small size of this group (three participants), it is hard to imagine how they might end up dominating governance arrangements in which the average attendance at meetings equals forty-four citizens.

The meeting-attendance data can also be used to explore the question of accountability from the perspective of location. Where do HFWC participants reside? Are they locals? Do they come from the surrounding region (Idaho, southern Montana, western Wyoming, northern Utah), or from other parts of the United States? Table 5.2 displays the results of this analysis. Surprisingly, there is more regional representation (39.4 percent) at HFWC meetings than local (34.4 percent), with "national" attendance scoring a respectable 23.9 percent of all participants. The high marks for regional representation clearly benefits from steady, relatively high rates of participation by state-level (Idaho) natural resource agency officials (13 percent of total participants). At the same time, however, it is somewhat remarkable, especially when compared to the Applegate Partnership situation, that 58 percent of the environmentalists participating in the HFWC live *outside the watershed*, with most being regionally based. Also of interest, and markedly different from the Applegate Partnership case, is the fact that almost half of all unaffiliated citizens live outside the watershed.

Staying True to Agency Mandates: Grizzly Bears and Road Closures

The HFWC, through the Watershed Integrity Review and Evaluation (WIRE) decision-making guidelines, pledges to obey all existing statutory obligations. Agency representatives from the state and federal levels interviewed for this project unanimously report that there has never been any pressure to stray from legal mandates in the five years of council operation, from 1994 to 1998 (personal interviews, 1998). An example illustrating the larger point involves the USFS and the decision to close forest access for the sake of grizzly bears. In this particular case, there is clear support for a national law (ESA), there is no conflict with either local or state laws, and there is clear support for a particular segment of the community (environmentalists). However, it can also be argued that accountability to the community and the state/region is simultaneously weakened given the zero-sum character of the outcome (environment *over* economy and community) and the fact that local and state representatives did not participate in the decision despite strong demands to do so.[2]

In the mid–1990s, as mandated under the National Forest Management Act (NFMA) of 1976, Targhee National Forest officials started to revise the forest plan governing agency goals for the use and protection of forest resources. USFS officials decided to release the draft plan and the environmental impact statement at a council meeting "because the broad cross-section of people in attendance . . . represent so many of the constituents and agencies we work with" (personal interview, 1998). One of the most controversial aspects of the draft plan involved the closure and dismantling of a significant number of forest roads to increase the amount of "connected" (unfragmented) habitat available to grizzly bears. The Forest Service believed that such action was not only warranted by the science associated with grizzly bear survival, but was also mandated by the ESA. Off-road motor vehicle enthusiasts and hunters mounted vigorous arguments against the closures because it would prevent them from utilizing large, historically accessible tracts of the Targhee. Yet the council was unable to come to a consensus regarding the revised forest plan primarily because the USFS refused to budge on the road-closure issue—an issue that the agency viewed as part of its legal mandate. There was even some discussion about whether to subject the road-closure decision to the Watershed Integrity Review process until it became clear to participants that this was not a decision for the council; rather it was a USFS decision under the National Forest Management Act of 1976.

Helping Agencies Do Their Thing: Small Projects, Endangered Species, and TMDLs

Wherever mission and goal compatibility exists between the HFWC and government agencies, the HFWC can and does sponsor cooperative programs and research important to watershed management and health. In this case, cooperation extends the effectiveness of existing government agencies by providing additional resources—financial, human, political, informational—that are then used to achieve agency missions and goals. Different projects, of course, spread accountability benefits in different patterns. Nonetheless, because resources are being coordinated and targeted to address *existing* agency mandates, collective accountability ends up being the norm even though there are varying combinations of levels (see table 5.4). For example, the HFWC has aided Fremont County officials in their attempts to control noxious weeds by identifying and locating such weeds, coordinating volunteer weed-pull efforts, and furnishing a total of $3,000 in funding between 1998 and 2000 (HFWC meeting minutes, June 20, 2000). Although in concert with the county (so community accountability), noxious weed control has become a major natural resource policy issue across the West in the last ten years and is the

Table 5.4
Linking outcomes to accountability: Henry's Fork Watershed Council

Outcomes	Individual (micro)	Community (meso)	State/region (mid-macro)	Nation (macro)
Helping Agencies Do Their Thing: Small Projects (varies by project)	• Strengthens/ clear support for *all* private landowners (noxious-weeds case) • Strengthens/ clear support for specific livelihoods and/or particular segments of the community (road-repair case; owners of motorized vehicles and recreation enthusiasts)	• Strengthens/ extends program reach (noxious-weeds case) • Strengthens/ supports positive-sum outcome (noxious-weeds, road repair, and frog/toad habitat enhancement cases; bioengineering) • Strengthens/ diverse array of participants • Representatives involved (noxious-weeds case)	• Preserves/no conflict with laws (bioengineering case) • Preserves/ clear support for laws (noxious-weeds case) • Strengthens/ extends program reach (noxious-weeds and Harriman Park bioengineering cases) • Representatives involved (noxious-weeds and Harriman Park bioengineering cases)	• Preserves/no conflict with laws • Strengthens/ extends program reach (USFS; road repair and frog/toad habitat enhancement cases) • Representatives involved (road repair and frog/toad habitat enhancement cases)
Helping Agencies Do Their Thing: Yellowstone Cutthroat Trout	• Strengthens/ approach increases sensitivity to individual situations	• Preserves/no conflict with laws • Strengthens/ supports positive-sum outcome • Strengthens/ diverse array of participants	• Preserves/no conflict with laws • Representatives involved	• Preserves/ clear support for laws (NFMA; ESA) • Strengthens/ extends program reach (USFS) • Representatives involved

Table 5.4 (continued)

Outcomes	Individual (micro)	Community (meso)	State/region (mid-macro)	Nation (macro)
Helping Agencies Do Their Thing: TMDLs and Custom Fits	• Strengthens/approach increases sensitivity to individual situations within context of the larger public goal (clean water)	• Preserves/no conflict with laws • Strengthens/supports positive-sum outcome • Strengthens/diverse array of participants • Strengthens/increases sensitivity to unique characteristics of "place"	• Strengthens/refines program application through new information, improved accuracy Idaho Department of Environmental Quality (IDEQ) • Representatives involved	• Preserves/clear support for laws (CWA) • Representatives involved

subject of Idaho state-level laws (state/region accountability). Moreover, individual-level accountability is enhanced because success on public lands also benefits private property interests, especially those in agriculture and ranching, by slowing the spread of weeds.[3]

The HFWC also has been integrally involved in road rehabilitation efforts throughout the watershed. It has assisted the USFS in repairing roads so as to minimize erosion and has helped Fremont County to adjust road levels and to design and place culverts along Sheridan Creek to better accommodate high water flows in the spring. In addition, the HFWC provided matching funds to help the USFS protect and enhance pond habitat along North Leigh Creek for the Western boreal toad and spotted frog, two species identified as sensitive by state and federal agencies (*Fremont County Herald-Chronicle* 1995b; personal interview, 1999). Further, the council assisted and encouraged Harriman State Park officials to employ, for the first time, a bioengineering solution as part of stream restoration on Sheridan Creek (Diversion 10). The choice of a nontraditional solution came about only after HFWC participants hired a private consultant to develop an alternative to the original Natural Resources and Conservation Service (NRCS) diversion design, then communicated the possibility of bioengineering to park officials (personal interview, 1999).[4]

The HFWC is also supporting proactive efforts to address the issue of declining stocks of Yellowstone cutthroat trout. Through the Cutthroat Trout subcommittee, the HFWC is comprehensively mapping existing fish populations, restoring habitat, and transplanting genetically compatible cutthroats into viable habitat. The goal is to stabilize and promote native cutthroat populations as well as to forestall or mitigate the potential listing of Yellowstone cutthroats under the federal ESA.

A key element of this work involves the Native Trout Inventory Project. The Forest Service, although required by law to map cutthroat populations on public lands, was unable to conduct the mandated review because of lack of funds and personnel. Together with the USFS, the HFF approached the HFWC seeking support for a proposal designed to inventory native cutthroats. The HFWC provided initial seed money for the project, the HFF provided the field crews, and the USFS provided a truck and a supervisor to coordinate the project. The inventory project surveyed almost 200 miles of streams in the Henry's Fork watershed, including stream reaches on Targhee National Forest property. In the Henry's Fork watershed (including Falls River), cutthroat were present in 20 of 136 stream reaches surveyed and isolated from nonnatives in 8 of these reaches. This represents occupancy in 16 percent of their historic range. In the Teton drainage alone, cutthroat were present in 16 of 32 streams surveyed and were the only trout present in 3 of these. This represents occupancy in 72 percent of all fish-bearing habitat, all of which falls within the specie's historic range (Jaeger, Van Kirk, and Kellogg 2000).

The collaboration on cutthroat trout intends to create communitywide benefits through strengthening biodiversity (by bringing back fish native to the area) and by staving off the new constraints on landowner decision making mandated by an ESA listing. An ESA listing would especially hamper the decision-making freedom of those with property adjacent to lakes and streams and those in agriculture dependent on irrigation water for their livelihood. At the same time, the project strengthens accountability to the macro-level (nation) by taking positive steps toward the ultimate goal of the ESA—to restore stocks of threatened and endangered species (see table 5.4).

Further, the HFWC creates "new" information and, in some cases, possesses more detailed and comprehensive information about resource conditions than the agencies responsible for environmental management. The additional, watershed-specific information can help agency decision makers make better decisions that are more likely to "fit" the actual on-the-ground conditions of the watershed (see table 5.4). For example, the Idaho Division of Environmental Quality (IDEQ) has extremely limited information regarding the water quality of the water bodies in the region, especially in areas surrounded by private lands, yet is mandated by court action to classify streams according to federal Clean Water Act Total Maximum Daily Load (TMDL) standards. The limited nature of the data means that, in most cases, IDEQ will set standards for major water bodies that automatically apply to adjoining tributaries. The HFWC fears this will lead to many water bodies being misclassified to the detriment of the resource and that the standards and ensuing management efforts will not fit the stream in question because they will either be too stringent or not stringent enough. Because the HFWC is the official state-chartered water-shed advisory group (WAG) for the Henry's Fork area, it is expected to give advice on water quality and habitat health issues directly connected to water quality to IDEQ and others. Pursuant to this expectation, the HFWC created the Water Quality subcommittee, on which any HFWC participant can serve.[5] The subcommittee's purpose is to make sure that when agencies make decisions they are informed with the best possible data. Toward this end the Water Quality subcommittee has assisted IDEQ with the upper Henry's Fork and Teton River subbasin assessments, using riparian habitat assessments completed by HFF and integrating USFS, BLM, and other agency data into a single document. The more comprehensive data essentially challenges IDEQ to come up with a classification scheme that more accurately captures the true diversity of stream conditions in the watershed.

Restoring Habitat and Watershed Connectivity: Sheridan Creek and Fish Ladders on the Buffalo River

The Sheridan Creek stream restoration project is the first major attempt to actively pursue the HFWC goal to identify, target, and reestablish the *connectivity* of tributary streams in the watershed. Connectivity is about making sure that tributary streams, such as Sheridan Creek, are physically connected (in terms of fish migration) to the Henry's Fork. Started in 1995, the specific goals of the Sheridan Creek project are to

• Restore the stream to its historical channel(s) and restore a natural flow regime,[6]
• Restore habitat in the river and along the streambank (spawning beds; vegetative cover),
• Improve water quality (water temperature especially), and
• Reconnect the natural stream channel to Island Park reservoir so that migrating fish can again access traditional spawning grounds above the lake. (Gregory 1997)

Key to the project are the redesign and rebuilding of ten different water diversion structures, and the drilling of wells away from the streambank as an alternative source of water for cattle. The expectation is that these steps will lead to less erosion and damage to streambanks, and therefore to greater opportunities for native vegetation to flourish.

The HFWC, by providing critical funding for the Sheridan Creek restoration project,[7] coordinating the resources of the many public and private stakeholders with some form of jurisdiction over or interest in the stream, administering the Environmental Protection Agency (EPA) grant, and providing a deliberative forum for forging agreement over the restoration plan,[8] demonstrates the broad benefits of a community-based collaborative approach to managing watershed resources (see table 5.5). The environment benefits, as do irrigators and farmers with control over the water rights. The new diversion structures help the FMID monitor stream flows with greater accuracy, while ranchers in the immediate vicinity end up with consistent delivery of their own water rights and subsidized reconstruction of their largely nonfunctional water diversion structures. Considerable benefits are also expected for areas downstream of the project. To the extent that cattle do not forage primarily on streambanks, water quality should improve. Finally, there are collective benefits for the state and nation levels of accountability. Improved habitat, cleaner water, and healthier fish runs address water quality, recreation (fishing), and, more broadly, environmental protection concerns expressed in many existing statutes. Participants in the HFWC, including several FMID board

Table 5.5
Linking outcomes to accountability: Henry's Fork Watershed Council

Outcomes	Individual (micro)	Community (meso)	State/region (mid-macro)	Nation (macro)
Restoring Habitat and Watershed Connectivity: Sheridan Creek	• Strengthens/clear support for specific livelihoods and/orparticular segments of the community (farmers)	• Preserves/no conflict with laws • Strengthens/supports positive sum outcome • Strengthens/diverse array of participants	• Strengthens/extends program reach (IDFG) • Representatives involved	• Preserves/clear support for laws (see text) • Representatives involved
Restoring Habitat and Watershed Connectivity: Buffalo River Fish Ladder	• Strengthens/clear support for specific livelihoods and/orparticular segments of the community (fishing; recreation)	• Preserves/no conflict with laws • Moderate/not entirely positive sum (some must pay more for electricity)	• Preserves/no conflict with laws • Strengthens/extends program reach (IDFG) • Representatives involved	• Preserves/clear support for laws (FERC) • Strengthens/extends program reach (USFWS) • Representatives involved

members, are excited about the prospect of restoring healthy fish runs to Sheridan Creek; they recall "how good the fishing used to be when [they] were young" (personal interviews, 1998 and 1999).

The HFWC also endorsed a consensus agreement between Buffalo Hydro, Inc., Idaho Department of Fish and Game (IDFG), the USFS, the U.S. Fish and Wildlife Service (USFWS), HFF, and others to add a fish ladder to Buffalo River Dam. The ladder reconnects the upper reaches of the Buffalo River, which had been closed to fish migration since 1938, to the larger Henry's Fork system. Although the Federal Energy Regulatory Commission (FERC) did not mandate a fish ladder when it relicensed the dam in 1994 (citing insufficient research), data produced by the HFF and IDFG—both key players in the HFWC—indicated that Buffalo Hydro's dam was blocking fish access to critical spawning and rearing habitats above the dam. Research shows that successful spawning in the Buffalo River could provide between 32,000 and 63,000 rainbow recruits (juvenile trout) and, eventually, up to 4,400 rainbow trout 16 inches or longer on an annual basis (Van Kirk and Giese 1999; Van Kirk and Beesley 1999).

In the end, not only was a state-of-the-art underwater monitoring system installed to type, measure, and count fish at the ladder, the entire cost of the ladder (estimated at $13,000) "was completely underwritten by Buffalo Hydro, Inc. on a voluntary basis" (Brown 1996a, 5–6). The cooperation on the Buffalo River extends beyond the successful completion of the fish ladder in 1996. HFF and IDFG have also agreed to review the data produced by the monitoring system, count trout redds (nests), install traps to assess juvenile recruitment success from spring spawning activity, and compare creel census data in order to assess catch and harvest rates for the Buffalo River and the Box Canyon of the Henry's Fork (Brown 1996a, 5–6).

The Buffalo River fish ladder clearly assists state-level efforts to manage and improve the fisheries resource, while indirectly helping *individuals* in the community with ties to the recreation/fishing economy (see table 5.5). At the same time, the hydropower capacity goals originally supported by FERC and regulations governing dam licensing continue to be met (macro level). The burden of community-level accountability is, however, not as strong in this case because community members buying electricity from Buffalo Hydro must absorb the costs of the fish ladder.

Dam Privatization: The Case of Island Park Reservoir

For the last several years the FMID has expressed interest in privatizing the Island Park Reservoir Dam on the Henry's Fork River (at the head of the Box Canyon). As with many dams across the western United States, the federal government currently owns the dam, while the U.S. Bureau of Reclamation (USBR), an agency in the Interior Department, regulates water flows. As a result, privatization, commonly called *title transfer,* requires an act of Congress and (likely) final approval by the Secretary of the Interior Department. The story of the potential Island Park Dam title transfer and the HFWC, just as in the cases of Sheridan Creek and the Buffalo River fish ladder, is about forcing additional community- and regional-level accountability into what previously has been a private (individual) realm (see table 5.6). This is the case because despite the fact

Table 5.6
Linking outcomes to accountability: Henry's Fork Watershed Council

Outcomes	Individual (micro)	Community (meso)	State/region (mid-macro)	Nation (macro)
Dam Privatization: Island Park Reservoir	• Weakens/ *less* support for specific livelihoods and/orparticular segments of the community (irrigation; farming)	• Preserves/no conflict with laws • Strengthens/ supports positive-sum outcome • Strengthens/ diverse array of participants	• Preserves/no conflict with laws • Representatives involved	• Weakens/ privatizes control of dam and water resources (but see the chapter conclusion) • Representatives involved
Using Research (Springs) to Help Manage Competing Goals	• Strengthens/ approach increases sensitivity to individual situations within context of the larger public goal (effective water resource management)	• Preserves/no conflict with laws • Strengthens/ supports positive-sum outcome • Strengthens/ diverse array of participants • Strengthens/ increases sensitivity to unique characteristics of "place"	• Preserves/ clear support for laws (healthy fisheries; recreation; effective water resource management) • Strengthens/ extends program reach (IDFG) • Representatives involved	• Preserves/no conflict with laws • Preserves/ clear support for laws (USBR) • Representatives involved

Note: IDFG = Idaho Department of Fish and Game; FERC = Federal Energy Regulatory Commission; USFWS = U.S. Fish and Wildlife Service; USBR = U.S. Bureau of Reclamation.

that the dam is a public facility in terms of ownership, in reality it is privately operated for the benefit of irrigators (farmers). As one HFWC participant and environmental advocate puts it: "Ninety percent of the time USBR does what irrigators tells it to given [irrigators'] control over water rights" (personal interview, 1999; see also Reisner 1986; Wahl 1989; Gottlieb and FitzSimmons 1991).

To most environmentalists, *privatization* or *market-based governance mechanisms* are "fighting words, something to be opposed at all costs" because they are code words for self-interested, special government by the few, an abdication of public responsibility to care for the environment (personal interview, 1998; Weber 1998, 149–151). HFWC participants, however, "did not want to take [such] a knee-jerk reaction to the privatization proposal; [according to an environmentalist,] we see it as an opportunity to engage in deliberations to come up with a better outcome environmentally, . . . [especially compared to the status quo. After all,] the Bureau [USBR] is not famous for enforcing environmental laws; they side with irrigators, often at the expense of the environment, the vast majority of the time" (personal interview, 1999).[9]

Some participants also see the title transfer as a way to create a new governing body for stream flows exiting the reservoir, with authority shared among three main community-based groups: FMID, environmentalists, and Fall River Rural Electric Cooperative (the hydropower company operating the Island Park Dam) (personal interviews, 1998 and 1999).[10] The dam operations committee would control when and how much water could be released from Island Park reservoir *outside of the irrigation season, not during it.* The existing water rights regime would dictate flows on an as-needed basis during the summer irrigation months.

Others point to the possibilities for more efficient management of water flows stemming from what everyone agrees, including FMID, is central to the title transfer proposal—a scientifically grounded river operations model (see the next section). As an independent scientist notes, this is the key "to knowing if you're actually getting an environmental benefit, or if more water can be released for fish or for hydropower production at any one time without harming others' [irrigators'] interests" (personal interview, 1998). An irrigator explains, "instead of having to spill water over the dam in the spring in high water years, we can release water early [during winter months] at a time when the fish get the most benefit" (personal interview, 1999). The added certainty is likely to translate into increased hydropower revenues (i.e., less spillage, more water through the power-generating turbines) that create additional opportunities for innovation. Some have proposed that the extra revenues be shared between the power company and the community in a watershed fund devoted to, for example, supporting irrigation-canal repairs, habitat improvement, more research, or electricity subsidies when the river operations model suggests it would be better to electrically pump ground water rather than draw water from surface flows.

Although these are all possibilities associated with the dam title transfer, the FMID almost short-circuited the deliberation and negotiation process when it announced to the March 1999 HFWC meeting that it was going to Congress with its own privatization proposal. FMID did not ask for any support or input from the council. Jan Brown of the HFF and others in the HFWC were shocked: "Everyone just went, "What?!" Brown, in particular, reacted immediately by playing hardball politics. She stood up in the very same meeting and announced that she was taking the next flight to Washington, D.C., to meet with national environmental groups and key members of Congress, and to introduce a competing bill in the appropriate committee. Yet the HFWC's rebuke of the FMID proposal was handled quietly. It was done without insulting FMID, without any news releases about what they did, and it was all aboveboard. Brown and others communicated to FMID that if they pulled their original proposal, joint negotiation over the privatization proposal was still a viable option (personal interviews, 1999).

Brown's trip to Washington got FMID's attention and reminded them of two things. First, other community members had the power to stop the passage of, or to seriously alter, FMID's preferred bill. Second, the end run around the HFWC jeopardized the effectiveness of a community-based institution and, by extension, the trust and good will that the FMID had so carefully nurtured for more than five years. FMID board members quickly realized that the decision was about much more than just dam privatization, it was about the kind of community they wanted to live in. And it was a test of their willingness to work with and listen to the concerns of others on a matter of critical importance to the watershed as a whole.

As a result, the FMID board of directors agreed to put their preferred legislation on hold in April 1999. They also agreed to work toward developing a scientifically based river-operations model and to negotiate the title transfer with environmentalists (HFF; Land and Water Conservation Fund of the Rockies[11]) and Fall River Rural Electric Cooperative. Once negotiated, the privatization proposal will be presented to the HFWC for its endorsement (personal interviews, 1999).

Using Research to Help Manage Competing Goals

The HFWC has financed key portions of the Henry's Fork Springs Research in cooperation with Utah State University, the Idaho National Engineering and Environmental Laboratory (INEEL), and a number of other federal and state entities, including the Idaho Department of Fish and Game (Benjamin 2000). In a typical winter, water flows are low out of Island Park Reservoir into the Box Canyon and Harriman Park sections of the Henry's Fork River, two of the primary river sections that give the Henry's Fork its reputation as a world-class rainbow trout fishery. Instead of the historical average natural winter flow rate of 450 cubic feet per second (cfs), the FMID and the USBR generally limit winter releases to 200 cfs or less (Van Kirk 1996, 9). The low flows exert a negative effect on the long-term health and survivability of rainbow trout (Benjamin and Van Kirk 1999; Mitro 1999). As a result, the HFWC would like to see higher winter flows to better protect the fishery, yet without infringing on the ability of irrigators to call on their water rights during the summer growing season. Success at managing these competing goals requires a flow regime that incorporates a better understanding of the sources of water flows as well as the rates and timing of flows *into* the reservoir.

The Springs Research Project is designed to clarify where Island Park Reservoir water originates by quantitatively specifying the relationship between snowmelt, groundwater, and surface flows, as well as the amount that each source contributes to the reservoir pool. To the extent that the research succeeds and a reliable source-flow model is developed, FMID has expressed a willingness to delay the annual full reservoir pool fill-date target from April 1 to May or June, thereby allowing the release of "extra" water from the reservoir during the winter to more closely mimic predam flows *if* the model indicates it can be done without harming required summer flows to individual farmers (personal interviews, 1998 and 1999).

The Springs research manages to keep macro-level accountability intact because the USBR still meets its water resource management obligations, although on a different time schedule, while simultaneously enhancing environmental protection (i.e., healthier habitat year round) (see table 5.6). By extension, community accountability improves because there are likely to be more, higher-quality recreational opportunities for locals and positive economic effects for the local recreation sector, yet *without* negatively affecting the ability of area farmers (individual accountability) to call on their legal allotments of water as needed. The new source-flow model also pays homage to state-level goals, particularly in the areas of fisheries health and recreation.

Never Take No for an Answer (Especially If the Problem Won't Go Away)

The entrepreneurial character of the HFWC is demonstrated by a case involving the Diamond D Ranch on Targhee Creek. The Targhee Creek case shows, once again, how HFWC outcomes adhere to the concept of broad accountability by simultaneously preserving or strengthening accountability at all four levels (see table 5.7).

The HFWC helped the Diamond D Ranch on Targhee Creek find funding to upgrade a canvas-and-plywood water diversion structure. The Watershed Council also alerted the rancher to the possibility of installing a more environmentally friendly, bioengineered solution for the same amount of money as a traditional structural (concrete) solution. The Targhee Creek rancher had originally worked with the federal NRCS to fix the diversion structure and allow cutthroat trout access to previously blocked upstream spawning areas. The NRCS suggested a concrete diversion structure that would also include a fish ladder, yet was unable to find cost-sharing money in its budget to facilitate construction. The ranch owner then approached the HFWC for funding assistance, in part because they had been instrumental in funding a prior project to improve stream habitat on his land. The HFWC demurred, however, given the fact that it had already invested a substantial sum in the earlier project and given its concern that HFWC moneys be spent on projects throughout the watershed.

Yet, rather than simply saying no and wishing him luck, the HFWC helped the rancher solve the funding puzzle. Dale Swenson, cofacilitator of the HFWC and executive director of the FMID, made the rancher aware of a nontraditional funding source for his project—a USBR program that provided a 50-50 cost share for projects improving water management. At the same time, some HFWC members, as well as the rancher,

Table 5.7
Linking outcomes to accountability: Henry's Fork Watershed Council

Outcomes	Individual (micro)	Community (meso)	State/region (mid-macro)	Nation (macro)
Never Take No for an Answer: Targhee Creek	• Strengthens/ clear support for an individual's problem (financial subsidy, innovative linking of resources for private landowner)	• Preserves/no conflict with laws • Strengthens/ supports positive-sum outcome (water quality and fish-migration benefits)	• Preserves/ clear support for laws (environmental quality) (IDEQ)	• Strengthens/ extends program reach (NRCS)
Changing Worldviews	• Weakens (people shift toward the idea of enlightened self-interest and a focus on win-win-win outcomes)	• Strengthens	• Strengthens	• Potentially strengthens, but unclear

were concerned that the placement and design (concrete) of the diversion structure—on a tight bend in the stream—risked failure during high-water conditions, and subsequent, potentially long-term damage to adjoining riparian areas. Wanting, at minimum, to give the landowner a choice of different solutions, and at best, to seize the opportunity for employing and demonstrating the benefits of a more environmentally friendly solution, the HFWC used bioengineering techniques to design a new solution for the same cost ($18,000). Large boulders placed at several intervals along the affected length of Targhee Creek would allow for water diversion and the gradual dissipation of stream-flow energy (slowed water flow) in a series of steps (drop-offs). The "slowed" flow would minimize streambank erosion and sediment flows, while simultaneously facilitating fish migration. In the end, USBR funded the rancher's request and the rancher chose the bioengineered solution (personal interviews, 1998 and 1999).

Changing the Worldviews of Individuals

Evidence drawn from interviews with HFWC participants suggests that the new governance arrangement is developing new relationships, "new" trust among citizens, and a heightened sense of collective purpose such that worldviews are being transformed. Moreover, the changes appear to be effecting an increase in commitment to broad-based accountability, at least with respect to the community and region (see table 5.7).

Deliberating together and cooperating with others on efforts providing communitywide benefits has led to new, positive working relationships within the watershed. According to one HFWC participant, an unaffiliated citizen, this is "absolutely critical to building trust within the community. . . . To the extent that we investigate and cooperatively pursue projects that help all watershed residents gain something, trust will follow" (personal interview, 1999). A representative of recreation interests finds that "the one-on-one interaction helps us to see each other as individuals, as decent human beings who care about their families, their neighborhoods, rather than as caricatures or adversaries that go by the name of 'farmer' or . . . 'developer' or 'environmentalist.' The trust that comes from working together helps us learn to communicate more openly and honestly with each other" (personal interview, 1998). When asked to assess whether their participation in watershed council proceedings had changed their willingness to trust others, especially those they may have viewed as adversaries prior to joining the HFWC, 89 percent (24 out of 27) responded in the affirmative.

Those unfamiliar with the HFWC dynamic often are not prepared for such open, frank discussion. A board member of the HFF, attending a meeting for the first time in the summer of 1999, shortly after the FMID attempt to submit their original privatization proposal without HFWC support, was "dumbfounded at the quality of relationships [between people]. . . . There was so much direct and honest communication going on. . . . No one was playing games" (personal interview, 1999). A citizen who has been involved from the start interprets the change in attitudes as follows: "Five years ago when the Council began, people were so cold to each other. It makes me happy to see how much trust has developed among us and how warm, friendly, and comfortable we are with each other now" (public remarks, November 1998 HFWC meeting).

There also is evidence that at least some participants are now more willing to think of their own individual situations as connected to, or an extension of, the larger whole. When asked whether participation in the watershed council has led them to give greater weight to how proposed actions will affect the world outside of the watershed community, fully one-third of those interviewed said yes (9 out of 27). Another roughly 37 percent of those interviewed claimed the willingness to consider the effect of proposed HFWC decisions on the outside world as a starting point. When asked whether participation has led them to give greater weight to the benefits of proposed actions for the watershed community, 40 percent answered yes, while almost 50 percent claimed community-mindedness as an original position. For example, a state agency official observes that

my world was a very small one. I would . . . deal with individual [agency] clients one by one. Looking out for, or thinking about the big picture, or the interests of others beyond this was not my responsibility. . . . And then I became involved with the HFWC and boom, my view started to broaden. At one meeting, me and my [departmental] colleagues had our eyes opened up. It was clear that we couldn't simply operate from a narrow [agency-only] perspective anymore. We realized it was a whole new ballgame [in which] . . . we needed to start involving a lot more people in our decisions rather than just going out and doing it by ourselves. (personal interview, 1998)

Conclusion

As in the case of the Applegate Partnership, the story told by the various outcomes produced by the HFWC suggests that broad, simultaneous accountability is the rule. Eight of the eleven outcomes either preserve or strengthen accountability at all four levels—individual, community, state/ region, and national. The "Changing Worldviews" case offers strong support as well, given that accountability is strengthened for the community and state levels, while simultaneously being weakened at the individual level. A tenth case, that of "Grizzly Bears and Roads," offers moderate support for the broad-based, simultaneous-accountability proposition at the community and state levels, while clearly matching the promise of staying true to existing mandates with performance at the macro-level.

Only in the "Dam Privatization" case might it be argued that a degree of accountability has been lost to the national level, given the transfer of regulatory control away from the USBR. However, placed in the context of the USBR's poor environmental record, the current reality of largely private control over river operations, the HFWC insistence on guaranteed environmental benefits as a primary condition for title transfer, and the potential for hydropower efficiency gains under the proposed transfer arrangements, it may be more plausible to argue just the opposite.

Moreover, the cases display support for the proposition that the council maintains an explicit focus on governance performance as an essential component of the accountability equation. The examples show little concern with who gets the credit and who does the work. Instead, there is a clear concern for action and actually achieving on-the-ground results consonant with the overall "environment, economy, and community" mission. At the same time, claims of industry domination and an increased propensity to roll back established environmental laws find no support in these cases. Diverse, balanced representation among environmentalists and business/commodity interests and strong, consistent participation by state and federal bureaucracies with stakes in the watershed are the rule rather than the exception. Likewise, the outcomes suggest enhanced support for existing local, state, and federal laws, particularly those focused on environmental protection.

The outcomes also suggest that not only does the council increase individuals' support for collective meso-level (community) goals, it strengthens the connection of the watershed community to the macro-level by firmly integrating state- and national-level representatives into place-based decision-making processes. The tighter linkages between levels of government do not, however, suggest that government agencies are in the process of being captured by local interests. There is no evidence that agency representatives are being forced to choose between agency goals and watershed council goals. Rather, the obverse appears to be true. The HFWC is helping agencies achieve their own preset *public* goals by using innovative institutional means and by allowing agencies to catalyze new resources external to the agencies, among other things.

6

Preserving and Restoring Natural Resources in a Pristine, Nature-Dependent Community: The Case of the Willapa Alliance

On a rocky bluff overlooking white-sand beaches just north of the town of Ilwaco, Washington, a bald eagle rises from the Pacific surf with a large salmon in its talons. Fighting for altitude, the eagle strains with every beat of its wings to reach its nest, 150 feet off the ground in a 200-year-old fir tree, in order to provide life-giving sustenance to its offspring. Back on the ground, citizens of the Willapa Basin are engaged in a similar fight to sustain communities and an ecosystem that has seen better days.

The Willapa is an area of exceptional biological productivity in its forests and waters. Conifers grow faster in the rain forests of the Willapa than almost anywhere else in the United States. The 80,000-acre bay is considered the cleanest large estuary in the continental United States. It is one of the most productive oyster-growing areas in the world. Like many other parts of the Northwestern United States, its salmon and steelhead runs are legendary, with annual historical runs totaling millions of fish (Hollander 1995b). Its pristine character, relative solitude, recreational opportunities, and sandy beaches attract tourists from throughout the Northwestern United States.

Willapa residents have long capitalized on nature's bounty. Livelihoods directly dependent on natural resources—farming, fishing, shellfish, logging, and tourism—have dominated, and continue to dominate, the local economy. It is impossible to escape the fact that the people of Willapa *need* a healthy ecosystem as a prerequisite for a healthy economy and a healthy community (Willapa Alliance, 1998d, 3). As an editorial in the region's largest newspaper, the Chinook Observer, notes:

We who live in this secluded corner of the Pacific Northwest are awfully lucky in some ways. We're surrounded by beautiful land and water that have escaped large-scale industrial development or pollution. That beauty and purity are our most valuable natural resources, just as rugged mountains and deep snow are for Aspen [Colorado] and as palm trees and clean beaches are for Hawaii. Maintaining that "Willapaness," the special ambiance around the bay, is the single best step we can take to encourage economic development (1996, 2).

The problem is that by the late 1980s the community and the surrounding ecosystem were under attack from various angles. Spartina— an invasive, nonnative cordgrass—was destroying mudflats essential to shellfish production. Wild salmon, steelhead, and sea-run cutthroat trout runs were in precipitous decline throughout the region. The decline in chum salmon, in particular, contributed to an explosion of burrowing shrimp, further destabilizing shellfish beds and lowering productivity. Years of logging weakened the ability of the steep Willapa hills to hold water, leading to erosion, large increases in waterborne sediment loads, and more severe flooding events. Adding to this, residents were concerned about the challenges to Willapa communities and possible restrictions on business as usual posed by the potential ESA listings of spotted owls, marbled murrelets, and salmon (Dan'l Markham, in U.S. Senate, 1997, 48; Hollander 1995a, 3).

The threats to community well-being spiraled "into conflicts in which real dialogue [was] minimized, suspicions [rose], information [was] distorted, and fear and anger [were] begin[ning to] become prevalent" (Willapa Alliance and Pacific County Economic Development Council, 1997, 1). The situation was ominous, though "the sky [had not] fallen on [the Willapa Basin] . . . at least not yet . . . not completely."[1] However, the lack of constructive dialogue and the increase in conflict around natural resource issues was enough to convince many of the need for new institutional arrangements to more effectively manage the interface between humans and nature (Allen 1992).

Founded in 1992, the new institutional arrangements—the Willapa Alliance—promised allegiance to a broad public interest. Environmentalists, industry representatives, local government officials, agency folks, tribal people, scientists, small-business owners, and unaffiliated citizens came together using a collaborative format to "find solutions to ensure a healthy future for their communities, [and to] understand that ecosystems, communities, economies and government are interdependent and must be in sync to achieve sustainability, that is, long-term community productivity and health" (Markham, in U.S. Senate, 1997, 47–48).

Institutional arrangements notwithstanding, how have the results produced by the Willapa Alliance stacked up against the idea of broad-based accountability?

A Multifaceted Effort to Restore Historic Runs of Anadromous Fish

Within the ecological fabric of the Willapa Basin, salmon are an indicator both of ecosystem health and of the effectiveness of natural resource management strategies. The problem is that anadromous fish runs in the Willapa are being negatively affected by a host of factors—loss of spawning grounds, stream blockages by culverts and huge logjams, and decreased genetic diversity due to the decades-long reliance on hatcheries. There is also the matter of reduced stream productivity due to the loss of nutrients provided by returning fish. After salmon spawn, they die. The carcasses are a critical source of biomass, or nutrients for over a hundred other plant and animal species, including juvenile salmon. [2] Yet the Willapa watershed, like virtually every other watershed in the Northwestern United States and Canada, no longer receives the ecological benefit from dead and rotting salmon carcasses as in times past. In part this is because many fish are harvested (caught). In part it is because habitat degradation contributes to declining salmon runs.

To address the complex characteristics of the fisheries problem, the Willapa Alliance has crafted a multifaceted Fisheries Recovery Program. Relying on locally fostered implementation efforts, the Fisheries Recovery Program "asserts that science-based management approaches [that] are sensitive to economic and social needs . . . in a coordinated and prioritized manner . . . are the only hope for reversing the precipitous downward spiral of Pacific Northwest salmonid populations" (Willapa Alliance, 1996, 1). The expectation is that the combination of science and local involvement will have positive long-term effects on the ecological and economic health of the Willapa region (Willapa Alliance, 1995; Willapa Alliance, 1996c).

The net result is that accountability is broad based. The fisheries governance effort is responsive not only to individuals such as gillnetters (fishers) and the community writ large (the focus on economic, ecological, and community health), but to the state/region and nation as well. The Fisheries Recovery Strategy partners with state- and federal-level representatives, and supports and strengthens the capacity of existing state and federal laws, such as the Washington Wild Salmon Policy, the Washington Endangered Species Act, and the U.S. ESA, to achieve their publicly *pre*defined goals. This is done in part by bringing new, more comprehensive information to the table and by prioritizing resource expenditures to get the most bang for the buck. It is also done by harnessing private-sector funds for public purposes, albeit in voluntary fashion rather than through taxation (see Tables 6.1 and 6.2).

A Comprehensive Fisheries Management Plan

The Willapa Alliance spearheads the Willapa Fisheries Recovery Strategy (WFRS), a multiplayer effort "to meet jointly developed local and regional goals" (Willapa Alliance; 1996b, 2) (see table 6.1). More specifically, the collaborative effort aims to create a basinwide, science-based, stakeholder-driven fisheries recovery plan that includes ecological, harvest, production, and socioeconomic recommendations. The WFRS team's mission is to develop a strategy for increasing and sustaining the populations of the five anadromous fish species native to Willapa Basin— chinook, coho, and chum salmon, steelhead, and sea-run cutthroat— through the following actions:

• Restoring and maintaining key ecosystem functions and processes fundamental to the biological productivity of the system

• Improving current fish propagation practices to allow the highest level of hatchery production compatible with the maintenance of natural productivity, and

• Improving methods for utilizing natural resources to secure the health and continued abundance of these resources while developing stability in the economic sectors which depend on them (Willapa Alliance, 1996, 1)

Phase One of the Fisheries Recovery Program is a draft WFRS report, first released in 1995, that uses empirically based information on historical and current salmon populations, numerous watershed characteristics (such as water temperatures and riparian vegetation), and the particulars of past and current resource management practices "to build sound management strategies . . . [capable of] providing both ecosystem and economic benefits for the region" (Willapa Alliance, 1996a, 1; Willapa Alliance, 1996c, 1). Phase Two presents technical recommendations for reaching recovery goals and initiates on-the-ground projects that continue data-collection efforts, seek to restore priority[3] fisheries habitat, and engage innovative fish population enhancement and fisheries harvest programs.

The alliance, in partnership with the Weyerhaeuser Corporation, Pacific County Conservation District, local farmers and gillnet fishers, the Willapa Bay Water Resources Coordinating Council, the Washington Department of Natural Resources, the National Marines Fisheries Services, the Chinook and Shoalwater Bay Indian nations, and others, has raised and administered more than $1.3 million in pursuit of WFRS goals (Willapa Alliance, 1995; Willapa Alliance, 1996c; Hollander 1995c; Wold 1995a).

Taking Action: Relieving the Pressure on Naturally Reproducing Runs of Chinook Salmon
The WFRS identifies the Willapa River, along with the North, Long Island, and Long Beach watersheds as priority areas for developing innovative harvest and fish propagation projects. A key step in this direction is the Willapa Sustainable Fisheries Harvest Program on the Willapa River, otherwise known as the selective hatchery release and harvest project (see table 6.1). Allen Lebovitz, the science director for the Willapa Alliance, is the project leader, in cooperation with local fishers (the Willapa Gillnetters Association), state-level representatives (Washington Department of Fish and Wildlife; Willapa Regional Enhancement Group), the city of South Bend, Washington, and concerned citizens from area businesses like the Boondocks Restaurant in South Bend (Willapa Alliance, 1996h).

Traditional hatchery practices have long been criticized for releasing hatchery fish into river habitat without regard for the existence or health of remaining wild (naturally occurring) fish stocks. The argument is that hatchery fish increase the competition for limited food sources and prime habitat, thus putting a strain on ecological resources and weakening wild fish stocks, which makes them more susceptible to disease. At the same time, given that hatchery operations are designed primarily for supporting harvest opportunities by sport and commercial fishers, the mixing of hatchery with wild fish increases the chances that fishers will catch and remove wild fish along with hatchery fish.[4]

In response to these concerns, the Alliance's sustainable-fisheries program identified a "terminal" fishery site on the Willapa River. The ultimate goal of the four-year program (1996–2000) was to enhance fishing opportunities and the survivability prospects for remaining populations of naturally spawning chinook by relieving harvest and hatchery pressure:

Table 6.1
Linking outcomes to accountability: Willapa Alliance

Outcomes	Individual (micro)	Community (meso)	State/region (mid-macro)	Nation (macro)
Fisheries: Management Plan	• Strengthens/clear support for specific livelihoods and/or particular segments of the community	• Preserves/no conflict with laws • Strengthens/supports positive-sum outcome • Strengthens/diverse array of participants	• Preserves/clear support for laws (see text) • Representatives involved	• Preserves/clear support for laws (see text) • Representatives involved
Fisheries: Sustainable Harvest Program	• Strengthens/clear support for specific livelihoods and/or particular segments of the community (those involved in gillnet fishing; recreational fishing interests)	• Preserves/clear support for goals (economic development) • Strengthens/supports positive-sum outcome • Strengthens/diverse array of participants	• Preserves/clear support for laws (fisheries policies) • Strengthens/extends program reach (hatcheries) • Representatives involved	• Preserves/clear support for laws (ESA)

"The goal was to catch all the fish . . . and derive the maximum sport and economic benefit from the fish, [while simultaneously] . . . reduc[ing] impacts to naturally spawning populations" (Willapa Alliance, 1996h, 1; also see Wold 1995a, 13). Two net pens—enclosures for fish that are located in the river and designed to simulate natural stream conditions—were filled with 200,000 juvenile chinook every year. Salmon are raised in the net pens for several months each year prior to being released. The idea was to use the net pens in such a way that the fish not only became acclimated to their home river, but developed a natural tendency to return to their home river at different times of the year than naturally spawning stocks (Willapa Alliance, 1998d, 32). In addition, all fish were given identifying marks (clipped adipose fin) and an internal coded wire tag for monitoring after release. The identifying tags are used to see how many fish returned to the location of the terminal fishery, the timing of the return, where the fish were caught, and if any moved upstream or to other areas to spawn.

Taking Action: Restoring Chum Salmon Stocks for the Sake of the Bay's Ecosystem

Historically, chum salmon has been the most abundant salmon in the Willapa Basin, with annual runs numbering close to a million fish. Yet over the past forty years chum stocks have experienced a sustained decline, with current counts showing Chum at approximately 20 percent of historical averages (Hollander 1995b, 14; Willapa Alliance, 1996g, 10–11).[5] The dramatic decline in chum salmon starting in the early 1960s is matched by a corresponding rise in the numbers of chinook and coho salmon in the Willapa.

The increases in chinook and coho stocks, however, are not part of any natural ecological cycle. The increases stem directly from state resource managers' decision to launch an aggressive hatchery program to propagate chinook and coho at the expense of chum. Willapa-area hatcheries now have the capacity to produce 20 million chinook and coho annually. The rationale behind the choice is economic. Because chum, or dog,[6] salmon are oilier than other salmon, they are not as commercially lucrative, typically commanding market prices that are one-half or less those for chinook and coho.[7] The low priority accorded to chum has been accompanied by a fisheries management regime that "expanded fishing on all salmon without taking lower numbers of Chum into account" (Hollander 1995e, 14). As a result, the harshest critics of the state's management efforts conclude that "the Chum were intentionally killed off to help the coho and chinook" (Marston 1997a, 7; personal interviews, 1998).

The declining numbers of chum salmon are also creating ecological problems for the mudflats of Willapa Bay. In recent years, the populations of two kinds of burrowing shrimp, mud and ghost, have soared to a level out of balance for the bay's ecology "because chum [one of their chief predators] isn't here to hold them down" (Hunt 1995a, 7).[8] The shrimp destabilize mudflats, destroying valuable oyster beds and crowding out native species by changing the character of the mudflats:

The shrimp tear up the usually stable mudflat as they burrow through, leaving ground so soft that you sink down to your waist in just a few minutes. It's not really a mudflat anymore when there is a real shrimp infestation. It's more like mud soup. Other things can't live there. Oysters, crabs, eelgrass—they all get pushed out as the mud surface turns to stew. (Hunt 1995a, 7)

To help reverse the decline of chum salmon and stanch the damage inflicted by burrowing shrimp, the Willapa Alliance operates a chum salmon net-pen project on the Nemah River. Started in 1994, the project raises and releases between 300,000 and 400,000 chum each year (Willapa Alliance, 1996e, 4; Willapa Alliance, 1997a, 3) (see table 6.1).

Taking Action: The Willapa Watershed Restoration Partnership Program

Established in fiscal year 1994–95, the Willapa Watershed Restoration Partnership Program (WRPP) is a key on-the-ground component of the larger WFRS effort (see table 6.2). The purpose of WRPP is to restore natural watershed functions, using historical conditions as a model, and to increase and improve habitat for salmonids. Of the eight primary watersheds in the Willapa Basin, four—the Bear, Naselle, Nemah, and Palix—are identified as priority restoration, conservation, and monitoring watersheds within the purview of WRPP (Willapa Alliance, 1995; Willapa Alliance, 1996c). The four watersheds still have relatively strong native fish populations, although significantly smaller than historical levels, and are missing certain key physical characteristics traditionally associated with healthy salmonid habitat. For example, the presence of large, woody debris in a river plays a significant role in establishing the physical structure of stream channels. Woody debris alters water flow and creates slower-moving pools and side channels, which trap sediment and gravel necessary for spawning and provide structural cover or protection from prey. However, prior management actions, including stream cleaning to remove woody debris, road-building and timber-harvesting practices that increase siltation (hence smothering salmon-spawning and insect habitat), and the placement of dikes and levees to capture and redirect stream flows, have all damaged the ecology of Willapa streams.

A WRPP program starts by gathering information for the baseline conditions of the river in question. Rivers are examined for in-stream structure (e.g., rocks; woody debris), stream-flow pattern and speed, fish and invertebrate populations, water temperatures, dissolved oxygen levels, and turbidity (amount of suspended sediment). A recovery plan specific to the river is then devised. The Bear River, for example, is home to all the salmon, steelhead, and cutthroat fish species native to the Willapa. The short-term plan to restore missing ecological functions includes placing large pieces of wood (i.e., trees) at strategic locations within the stream channel to retain sediment, gravel, and other organic materials of nutritional value to fish and to increase protective cover for juvenile salmon. Long term, the plan focuses on increasing the growth rate, number and diversity of native fir trees in the riparian zone in order to provide shade, thereby lowering water temperatures and stabilizing streambanks (Willapa Alliance, 1996a).

In another case on the Willapa River, Weyerhaeuser Corporation is working together with the Washington State Department of Natural Resources and the Willapa Alliance on a $1 million project to fill in old roadbeds with gravel to restore hillside stability and minimize soil erosion. Sediment loading is "probably the watershed's greatest problem [and] roads are a major sediment source" (Willapa Alliance, 1998d, 31). According to Dan'l Markham and Dick Wilson, both longtime participants in the Alliance, it is an example of restoration activities fostered by collaboration and grounded in the kind of enlightened self-interest necessary for the alliance's vision for sustainability to succeed. Everyone, including Weyerhaeuser, admits that the reason such restoration work is needed is because of "a half-century of erosion-causing logging practices—especially road-building—that silted up the Willapa's streams" (Hollander, 1995c, 9).[9] Yet, as Markham argues, the more important

Table 6.2
Linking outcomes to accountability: Willapa Alliance

Outcomes	Individual (micro)	Community (meso)	State/region (mid-macro)	Nation (macro)
Fisheries: Restoring a Less Desirable Species of Salmon (Chum)	• Strengthens/clear support for specific livelihoods and/or particular segments of the community (commercial fishing; shellfish interests)	• Preserves/no conflict with laws • Strengthens/supports positive-sum outcome • Strengthens/diverse array of participants • Strengthens/focuses on unaddressed community-based problem	• Preserves/clear support for laws (Wild Salmonid Policy)	• Preserves/no conflict with laws
Fisheries: Willapa Watershed Restoration Partnership Program	• Moderate, indirect support for specific livelihoods and/or particular segments of the community (but of secondary concern)	• Preserves/no conflict with laws • Strengthens/supports positive-sum outcome • Strengthens/diverse array of participants	• Preserves/clear support for laws (water quality regulations (WDOE) • Strengthens/extends program reach (WDFW) • Representatives involved	• Preserves/clear support for laws (ESA, chinook and steelhead stocks; Clean Water Act, water quality)

| Fisheries: Taking Action on Private Land | • Weakens/private sector voluntarily funds projects benefiting community and region | • Preserves/no conflict with laws
• Strengthens/supports positive-sum outcome | • Preserves/clear support for laws (water quality regulations) (WDOE)
• Strengthens/extends program reach to private land (WDFW; WDNR)
• Representatives involved | • Preserves/clear support for laws (ESA, chinook and steelhead stocks; Clean Water Act, water quality) |

Note: WDWF = Washington Department of Fish and Wildlife; WDOE = Washington Department of Ecology; WDNR = Washington Department of Natural Resources.

point is that the timber giant is actually funding and doing "good environmental work" that not only corrects past mistakes, but enhances the chances that "successive generations will have a healthy environment to give them a good economic base" (Hollander 1995c, 9).

Taking Action on Private Land: Restoring Hillsides and Stream Connectivity

One of the largest timber companies in the world, the Weyerhaeuser Corporation owns two-thirds of the land in the Willapa Basin. As a result, participants in the Willapa Alliance accept that they have to live with Weyerhaeuser, even Allen Lebovitz, raised in the East, a biologist by training, a graduate of the Yale School of Forestry, and science director for the alliance. Confronted by a reporter with the charge that the alliance is too timid in its dealings with Weyerhaeuser, Lebovitz responds:

What happens when you take on the 1000-pound gorilla . . . by heads-on confrontation? You get slapped around pretty vigorously. They [Weyerhaeuser] own the majority of the watershed . . . outright. Lock, stock, and barrel. If you want to work with private industry/property . . . you do need to play by some of their rules. At the same time, that doesn't keep you from trying to win. (Manning 1997, 9)

Besides, from the perspective of alliance participants, "Weyerhaeuser has been both more responsive and progressive than the other logging companies and the state in trying to undo some of logging's damage" (Manning 1997, 9; Hollander 1995c; personal interviews, 1998). One example involving road restoration has already been noted. Weyerhaeuser has also been proactive in experimenting with the ecological benefits that come from reintroducing salmon carcasses to streams (Hollander 1995b; Manning 1997) and in implementing and funding projects that help restore stream connectivity and reduce sediment loads (Wold 1995b).

In the latter case, Weyerhaeuser partnered with the Willapa Alliance, the Washington Department of Natural Resources (DNR), and the WDFW on a $423,000 watershed restoration project on Weyerhaeuser property just east of Raymond, Washington (see table 6.2). Fish seeking to migrate to the upper reaches of the stream in question were blocked by a series of culverts that were either "hanging too high" or damaged in some way. In other areas, fish migration was hampered by the lack of clear passage across eroded ground, often in the area of logging roads. To reconnect the main channel to its upper reaches and to other tributaries, including wetlands in one case, the project installed forty-three culverts.

At the same time, the partnership focused on various "roads that were falling down the steep, canyonlike Willapa hills" and on adding large amounts of sediment to local streams (Wold 1995a, 14). The effort stabilized existing roads (4.54 miles), recreated historical drainage patterns in areas of mudslides and road blowouts, and abandoned other old logging roads (11.17 miles) by restoring the original hillside slope, then reseeding the area with grass and clover (Wold 1995a).

Fighting the Invasion of Spartina

Spartina is a species of cordgrass native to the U.S. East Coast. However, it is not native to western Washington's naturally occurring mudflats. In fact, it is considered one of the most dangerous invasive or noxious species in the Northwest, wreaking economic and ecological havoc as it spreads across Willapa Bay and Puget Sound. It chokes natural mudflats by trapping sediments, raising the elevation of tidal flats, and replacing mudflats with meadows. To date, spartina has already infested more than 10,000 acres of former mudflats on Willapa Bay (roughly 21 percent of all mudflats) and hundreds of acres in Puget Sound. Moreover, it is spreading at an approximate rate of 20 percent annually, because its traditional competitors, diseases, and predators are missing in the West Coast ecosystems. Yet funds devoted to the eradication of spartina are not sufficient to either slow or reverse its spread (Christensen 2000; Wilhelm 2000, 3).

Invasive, nonnative species damage ecological systems. Miranda Wecker, marine program manager for the Olympic Natural Resources Center at the University of Washington, argues that "the threat that [invasive species] pose to the perpetuation of native species and the very structure of native systems is staggering . . . , and it's just starting to be realized that not all natural processes are benevolent" (as quoted in Wilhelm 2000, 2). Wilhelm (2000, 2) notes that some "studies . . . conclude that 40 percent of the creatures on the endangered and threatened species list are there because of competition with invasive species" (?). In the specific case of Willapa Bay, Spartina destroys habitat—intertidal zones—vital to shorebirds migrating along the Pacific Flyway and crowds out primary food sources (e.g., eelgrass). Spartina also kills natural shellfish habitat by replacing the natural mudflats conducive to rapid oyster and clam growth with meadows inhospitable to shellfish.

In economic terms, invasive species of all kinds are becoming a major issue in the United States, where the estimated annual cost to the U.S. economy is $123 billion, and elsewhere as the massive growth in transportation and international trade spread these life forms around the globe (Wilhelm 2002, 1). More specific to the Willapa, spartina directly affects the economy by taking natural shellfish habitat out of production. Robin Downey, executive director of the Pacific Coast Shellfish Growers, located in Olympia, Washington, calls spartina "a huge threat" to Washington's shellfish industry, which generates $73 million in annual revenues and employs approximately 2,000 people. While many in the oyster industry agree with Downey's assessment, there is some disagreement over the degree of the threat posed by Spartina to shellfish productivity. Dick Sheldon, a local Willapa oyster grower, points out that "spartina infects the higher, least valuable seed land. So to say that spartina [eradication] is a thing that the oystermen are doing for self-interest is not [entirely] true. . . . [I'm] doing this for future generations and because [I] care about the bay" (Christensen 2000, 4; personal interviews, 1998).

Spartina also poses an economic threat due to the increased risk of flooding. The higher elevation of the former mudflats (now grasslands) changes drainage patterns and hampers the ability to shed winter rainwater, thereby increasing the risk of flooding for towns located along Willapa rivers, like Raymond and South Bend (Wilhelm 2000, 3).

The Willapa Alliance has been in the forefront of the movement to eradicate and control spartina. In the fall of 1994, the alliance coordinated and encouraged a number of different organizations to coalesce into a single community-based organization—the Spartina Coordinating Action Group (COAG). The Willapa Alliance and COAG then worked together to develop a more comprehensive, science-based picture of spartina's effects on Willapa Bay. The information was presented to the Washington State Senate majority leader, Sid Snyder, as well as to the State Senate Natural Resources Committee, and was instrumental in the passage of state legislation that provides funding[10] and a permitting process for managing spartina eradication efforts (personal interviews 1998; Willapa Alliance, 1995, 3–4).

The spartina eradication efforts of the Willapa Alliance demonstrate clear accountability to individuals in the shellfish industry (economy), but also a sensitivity to community concerns centered on the potential economic and human costs due to increased flooding and the health of the bay ecosystem itself (see table 6.3). Accountability to the state/region is also evident through proactive efforts to recognize and address the seriousness of the spartina problem, which is not simply a Willapa-specific issue. By taking steps to slow the spread of spartina in Willapa Bay, there is less chance that—with all its economic and ecological costs, it will spread to other parts of Washington State or to the Oregon coast.

Monitoring the Health of the Community, Ecosystem, and Economy

Early in the life of the Willapa Alliance, board members and staff decided that promoting the principles and practices of sustainable economic development would not be possible without a system of community-based indicators capable of charting progress toward the alliance's goals (personal interviews, 1998). According to the alliance, "indicators are like navigational aids, giving us points of reference to chart a more certain direction for our community" (Willapa Alliance, 1998d, 3). They "establish the fundamental linkages [among] science, education, conservation based development, and natural resource-based management . . . [by] gather[ing] and summariz[ing] information about the whole community of the Willapa Bay watershed: its natural residents [i.e., ecology and wildlife], its social settings, and its economic life" (Willapa Alliance, 1995, 4). They are an invitation to civic involvement and discussion by residents about how to take stock of the Willapa community, choose common goals, clarify choices, and measure progress toward common goals (Willapa Alliance, 1995, 4; Willapa Alliance, 1996g, 5). And, as an added benefit, indicators are a new tool that anyone can use to hold either the alliance or other units of government to account for their promises and programs (see table 6.3).

Published on a biennial basis starting in 1996, the *Willapa Indicators for a Sustainable Community* (WISC) gathers and centralizes data on dozens of indicators from more than seventy-five government, private, and

Table 6.3
Linking outcomes to accountability: Willapa Alliance

Outcomes	Individual (micro)	Community (meso)	State/region (mid-macro)	Nation (macro)
Fighting Spartina	• Strengthens/clear support for specific livelihoods and/or particular segments of the community (shellfish interests)	• Strengthens/extends program reach for fighting a major local problem • Strengthens/supports positive-sum outcome • Strengthens/diverse array of participants	• Preserves/clear support for laws • Strengthens/extends program reach • Representatives involved • Strengthens/efforts address a major regional problem	• Preserves/no conflict with laws
Monitoring and Transparency: Willapa Indicators for a Sustainable Community (WISC)	• Weakens/focus is on collective health of community • Weakens/increased transparency makes it harder to promote zero-sum outcomes	• Strengthens/new data increases general capacity of local government units • Strengthens/supports positive-sum outcome • Strengthens/diverse array of participants • Strengthens/increases transparency and ability to monitor government/governance decisions	• Preserves/no conflict with laws • Strengthens/increases transparency and ability to monitor government/governance decisions	• Preserves/no conflict with laws • Strengthens/increases transparency and ability to monitor government/governance decisions

nongovernmental organizations on past trends and present conditions. The report uses health as an integrating concept for monitoring the status of, and trends in, indicators related to the community, ecosystem, and economy. Indicators reflect the three major dimensions of health—productivity, resilience, and diversity—and "each suggests tangible ways that the health of the environment is tied to the vitality of the local economy and to community well-being" (Willapa Alliance, 1996g, 5). The integrated view of the Willapa is believed necessary because single indicators, or sets of indicators dedicated to only one of the three main policy concerns (environment, community, and economy), are too narrow in scope to be sufficient gauges of health, hence sustainability (personal interviews, 1998; Willapa Alliance, 1996g, 6).

The 1996 WISC report contains eleven primary indicator categories that fit under the broader rubrics of environment, economy, and community. Within the broader category of environment, there are three major indicator categories—water resource quality, land use patterns, and species viability. For example, in the water quality section, WISC readily admits that "there is no simple way to assess water resource quality locally or regionally" unless an extensive water quality monitoring program can be established (Willapa Alliance, 1996g, 6). However, given that no organization has stepped forward, government or otherwise, to fund such a comprehensive initiative, the alliance has settled on two primary indicators for assessing water quality. The first is derived from Washington Department of Fish and Wildlife data on the condition of oysters in the bay because "oysters are highly sensitive to water quality and nutrient availability" (Willapa Alliance, 1996g, 7). The second is more traditional—fecal coliform—a toxic bacterium found in human and animal waste. The condition of oysters has experienced a slow and steady decline of approximately 25 percent from the 1960s to the early 1990s, while the fecal coliform count has "held fairly steady since 1988" (Willapa Alliance, 1996g, 7–8). The decline in the oyster-condition index is cause for alarm; it "sounds a warning that water quality in the bay deserves closer scrutiny in terms of understanding and managing the causes of the decline" (Willapa Alliance, 1996g, 7).

The economy indicator category takes into account the productivity of natural resources, the availability of employment opportunities,[11] measures for economic or sectoral diversity,[12] and the equity effects of the economy on individual Willapa citizens.[13] The productivity of natural resources is measured in two ways, in terms of the economic value added by such resources and in terms of the harvest levels of key natural resources—timber, cranberries, oysters, and dairy cattle. For example, by the early 1990s, the value added by timber to the local economy had declined to roughly half the level of the late 1970s and early 1980s, but was twice what it was during the mid–1980s amidst a broad decline in U.S. housing starts. In terms of natural resource harvest levels, salmon and oyster harvests were lower during the early 1990s than previously, while cattle production (heads of dairy cows) has remained nearly level, and cranberry production has steadily increased (Willapa Alliance, 1996g, 12–13).

In the area of community there are four indicator categories—lifelong learning opportunities,[14] human health (e.g., percent of healthy-birthweight babies; infant mortality rates), citizenship (e.g., voter turnout; diversity of community organizations; crime rates), and stewardship. The stewardship indicator category tries to gauge whether community members are sensitive to the impacts of their lifestyle decisions on the surrounding environment. The alliance thus collects data on solid-waste management practices (recycling vs. landfill disposal), the health of the human-waste—septic and sewers—management system (e.g., reports of leaking septic systems), and household energy and water conservation efforts (see Willapa Alliance, 1996g, 25–26).

Using feedback from the community as well as stakeholders in state and federal agencies, the Willapa Alliance issued a second WISC report in 1998 (Willapa Alliance, 1998d). The 1998 report further refines the original list of indicators and develops more comprehensive information on the status of and trends in the Willapa environment, community, and economy.

Economic Development without "Fouling the Nest"

According to Michael Dickerson, director of marketing for the Shore-Trust Trading Group, of Ilwaco, Washington, "the majority of people [in the Willapa] really do care about the ecosystem and have been good stewards" of the environment (Hunt 1995c, 13). Many alliance participants, including a logger, an environmentalist, a Native American, an

elected state official, a state agency official, and a business owner, agree
that a good portion of area residents intuitively recognize the interconnectedness of economy and environment and the value of sustainable development (personal interviews, 1998). Residents appear to understand
that

> if one piece of this interlocking ecosystem structure is overused, poisoned, or de-
> stroyed, other pieces suffer. . . . [Residents] agree with scientists' image of an
> ecosystem: It's like an airplane in flight. Allowing too much destruction of ani-
> mals, plants, soils and other natural resources is like pulling rivets out of the wing.
> At some point, enough rivets will be pulled that the plane will crash. So, too, will
> the bay's economy. (Allen 1992, 1A)

Nonetheless, as in so many communities, there has always been a short-
age of capital designed specifically to promote sustainable-development
activities in the local economy. The gap between citizens' good intentions
and a broad-based sustainable community created an opportunity for the
Willapa Alliance to demonstrate the seriousness of its commitment to the
tripartite "economy, environment, and community" mission. The Alliance, along with Ecotrust, convinced the Southshore Bank—a Chicago-
based lending institution that focuses most of its efforts on community
development in poor Chicago neighborhoods—to set up shop in Ilwaco
under the name ShoreTrust Trading Group.[15] The purpose, according to
Spencer Beebe of Ecotrust, was to get "right out in the communities with
the bakers, the fishers, oystermen, crabbers, timber guys and cranberry
growers. The federal government lists the owl or some salmon run and
sends out $50 million worth of welfare checks two years later. We are
trying to dig in and find alternatives" (as quoted in Blumenthal 1997,
B13).

In 1993, the ShoreTrust Trading Group established a $3.5 million re-
volving loan fund to finance environmentally compatible, natural resource–based businesses.[16] It has also helped entrepreneurs put together
business plans and helped them find new niche markets for sustainably
produced goods (Hunt 1995c, 11). According to John Berdes, Shore-
Trust's managing director, "the heart and soul of this organization, rather
than making loans, is finding markets for products that come from here.
If we let the consumer know about this place and the ways in which
sustainability is being sought, there are advantages in the marketplace
we can use" (as quoted in Hunt 1995b, 12).

The key is to develop environmentally friendly businesses that add *local* value to products, rather than having businesses in the Willapa serve only as a source of raw materials/resources, a scenario in which much of the profit ends up outside the community (see table 6.4). For example, "raw fish brings in a dollar, processed fish brings in a better dollar. Part of that dollar is wages, part is profit that gets reinvested in the community locally" (Berdes, as quoted in Hunt 1995c, 12). A case in point is Josephson's, a Willapa area business. In 1995, one could either sell salmon to processors for 83 cents per pound or, as Josephson's does, one could smoke or otherwise process the same amount of salmon and sell it for up to $32 per pound or more (Hunt 1995c). ShoreTrust has also helped the Willapa Alliance devise a pilot marketing effort for the Willapa Sustainable Fisheries Harvest Program. The idea is that "the sustainable way in which these fish are reared and harvested, . . . can command a higher price in the market, returning a greater economic benefit to the fishers who harvest them" (Willapa Alliance, 1996h, 1).

To date, the ShoreTrust Trading Group has made more than fifty loans focused on three primary sectors—red alder wood; nontimber forest products such as shiitake mushrooms, cones, cascara bark, and ferns; and seafood. One loan supported a local company's idea for using seafood waste—shrimp and crab shells—to extract chitosan, a material used in dissolvable surgical sutures and some cosmetic products. Another effort helped local fishers in Willapa sell salmon for extra profit 170 miles away in Ballard, Washington. The loan capital allowed them to purchase the equipment necessary for icing down the fish and getting them quickly to a market willing to pay $3 per pound rather than the competing 85-cent-per-pound-price in local markets (Blumenthal 1997; Hunt 1995b).

Nor has ShoreTrust limited investments to the Willapa area. They have financed the cleanup of an old mill site in Astoria, Oregon (just south of the Columbia River's mouth). And they have provided funding for a red alder furniture business in Oregon's Willamette Valley. Red alder is a type of hardwood with a relatively small-diameter trunk. As a result, it is typically treated as a "weed tree" and has been underutilized in the past. Yet red alder is not only native to the Willapa and the surrounding region, it also adds ecological value by pumping nitrogen into the soil, thereby "healing the soil and improving soil quality." In addition, it can be harvested in a manner—thinning—that leaves a relatively limited

Table 6.4
Linking outcomes to accountability: Willapa Alliance

Outcomes	Individual (micro)	Community (meso)	State/region (mid-macro)	Nation (macro)
Economic Development without "Fouling the Nest"	• Strengthens/clear support for individual business owners • Strengthens/limited array of interests involved (banks and businesses)	• Preserves/clear support for goals (economic development) • Strengthens/supports positive-sum outcomes	• Preserves/no conflict with laws	• Preserves/no conflict with laws
Diversity of Representation	• Difficult to assess given lack of records (see text)	• Strong/ broad cross-section of community-based stakeholders, including Native Americans	• Moderate involvement of state-level representatives	• Generally weak involvement of national-level representatives
Changing Worldviews	• Weakens (people shift toward the idea of enlightened self-interest and a focus on win-win outcomes)	• Strengthens	• Strengthens	• Potentially strengthens, but unclear

ecological imprint during harvesting, leaves very little waste, and promotes regeneration of existing red alder stands (Blumenthal 1997; Hunt 1995b, 12).

Getting a Diverse Public Involved

As in the other two cases, inclusiveness (diversity) is an integral part of the Willapa Alliance's approach to governance. In the words of one environmentalist, it "affects everything we [in the alliance] do. In fact, you would be hard pressed to find one project that we have done that does not have multiple numbers of diverse people and organizations involved" (personal interview, 1998). Yet, in some respects, the overall case for inclusiveness as it relates to levels of accountability is weaker than in the HFWC and Applegate Partnership situations, given fewer opportunities for involvement and the heavy focus on community as well as the individuals within the Willapa Basin. Of the five efforts, the leadership summits can plausibly be argued as providing accountability to the state, although it appears to be a relatively weak relationship. The extensive outreach by the Willapa Fisheries Recovery Program, on the other hand, suggests stronger accountability links to both the state and national levels. At the same time, a caveat is in order: the general lack of record-keeping in terms of names and interest affiliations for the alliance's public meetings means that a complete "diversity" analysis simply is not possible. For these reasons, the discussion of the alliance is more general than that of the other two cases and the conclusions less certain, except that it is clear there is strong accountability to the community level (see table 6.4).

First, the board typically includes a broad cross-section of community interests including timber interests (both large corporate farms and small tree lot owners), farmers, environmentalists, the Shoalwater Bay Tribe, recreation and tourism interests, fishing interests, and Ecotrust. There are also usually several other community members, or concerned citizens unaffiliated with any organized interest (Hollander 1995a, 3; Willapa Alliance, 1995). Seats on the Alliance board are not reserved for specific types of interests. Instead there is an informal agreement that when a member of a particular interest category leaves the board a concerted attempt will be made to fill it with another community member from the same category. Potential new board members are nominated and voted on by current members. By 1997, the board contained only nine of the seventeen original founders (Willapa Alliance, 1997a, 2). In general, board meetings are not open to the public in the same way as in the other two cases of GREM. The public is welcome, but virtually all of the discussion occurs among Board members.

Second, the Willapa Fisheries Recovery Program (WFRP) is a good example of how the alliance's cooperative partnerships often reach across levels of governance to include a wide variety of stakeholders. The WFRP involves twenty-eight cooperating organizations, seven philanthropic foundations,[17] individual local fishers, and other private citizens and businesses. The effort includes four federal agencies (EPA, National Marine Fisheries Service, Natural Resource Conservation Service, and the U.S. Fish and Wildlife Service), three state agencies (Washington Departments Of Ecology, Natural Resources, Fish and Wildlife), two state-level conservation groups (Washington Trout, Oregon Trout), two national environmental advocacy groups (the Nature Conservancy, Trout Unlimited), and three local administrative units.[18] Native Americans are also participating through the Chinook and Shoalwater Bay tribes, as are local elected officials (Pacific County Commissioners) (Willapa Alliance, 1996, 2).

Third, since the spring of 1996, the alliance has hosted a biennial Willapa Indicators Leadership Summit. The conference's purpose is to invite the people of Willapa to study and critique the biennially issued *Willapa Indicators for a Sustainable Community* report as a way of discovering "common understanding and mutual solutions . . . for long term community well-being" (Willapa Alliance, 1998c, 1). More than 100 people attended each of the first two Summits, including the editor of the region's largest community newspaper (the *Chinook Observer*), former Washington State Governor and Ambassador to Switzerland Booth Gardner, the Washington State Senate Majority Leader Sid Snyder (D), state-level administrators from natural resource agencies, business leaders, agricultural interests, and more. Working groups at the daylong summit are asked to make recommendations linked to the idea of a sustainable community. Major themes coming out of group discussions have included, among others, recommendations to refine and expand the WISC project, and to find ways to "care for the environment while providing family wage jobs" (*The Chinook Observer* 1996; Willapa Alliance, 1996b; Willapa Alliance, 1998c, 1).

Fourth, the alliance organizes and supports the Willapa Science Advisory Group (WSAG), a mixed group of local governmental and nongovernmental scientists, educators, and researchers. The WSAG makes annual reports to the public each spring using an all-day format.

Finally, in 1998 and 1999, the alliance took more concerted, proactive steps to make more people, particularly elected officials and leaders in the business community, familiar with the WISC and the *Nature at Home* resource guide: "We sat down with [Pacific] county commissioners and city council people. We met with chambers of commerce. We took along the indicators report and said, 'Here's what this means to us, what does it mean to you?' We essentially challenged people to begin integrating this new information into their own plans and programming" (personal interview, 1998).

Transforming Worldviews

Members of the Willapa Alliance see their collaborative, participative efforts to improve the Willapa Basin as "a visionary project . . . [aimed at promoting a] paradigm shift in the way the whole community understands itself" (Hollander 1995c, 8–9; Marston 1997c, 1; personal interviews, 1998). The expectation is that the institutional changes will redefine the constraints on behavior, opportunities for action (benefit), and the ground rules for social and political interaction. The changes in the community dynamic are expected to lead to changes or transformations in behavior as citizens learn to reconceptualize the relationship between environment, economy, and community and their own individual relationship to the collective whole. Evidence drawn from interviews with alliance participants suggests that new, constructive, trust-based working relationships are developing among participants and different elements of the community. The interviews also suggest that the collaborative, participative alliance framework helps to foster or, in other cases, reinforce a heightened sense of collective purpose in individual participants that can be interpreted as an increase in commitment to broad-based accountability to the community and regional levels (see table 6.4).

Willapa Alliance participants were asked to assess whether their participation in alliance proceedings had changed their willingness to trust others, especially those they may have viewed as adversaries prior to joining the alliance. In nearly every instance (20 out of 21 interviews) the answer was yes, participants now trust others, even adversaries, more than before. An environmentalist-participant from a state-level group believes that the alliance "has done a good job of building trust, especially between environmentalists and commodity-based interests. . . . [And] the increase in trust has been good ecologically because [business participants] are more open to hearing about how their practices affect the environment" (personal interview, 1998). According to other participants, including a scientist, two loggers, and an environmentalist," one of the things that the alliance has done is to fill an [institutional] niche in the Willapa as a trust-builder. It builds the kind of trust necessary for consensus building within the community between those of different views. . . . The level of trust has only been developed because the alliance does good things for the community through its accomplishments and achievements as well as just [establishing] a working relationship with different parts of the community. . . . There is a grander spirit of cooperation now" (personal interviews. 1998).

In addition, at least some participants are now more willing to view preferred outcomes from a narrow, self-interested perspective and are more disposed to believe that their individual circumtances are inextricably linked to the the larger whole. When asked whether participation in the alliance has led them to give greater weight to how proposed actions will affect *the world outside of the watershed community,* roughly 29 percent of those interviewed said yes (6 out of 21). Another 33 percent of those interviewed claimed the willingness to consider the effect of proposed alliance decisions on the outside world *as a starting point* (i.e., the institutional dynamic matched or reinforced their original position). When asked whether participation has led them to give greater weight to the benefits of proposed actions for the *watershed community,* 52 percent answered yes (11 out of 21), while 38 percent claimed community-mindedness as an original position (8 out of 21).

Conclusion

Participants in the Willapa Alliance are redefining and broadening the idea of policy or outcome success to include an explicit focus on the question of accountability "to whom" in addition to the more traditional focus on accountability "for what." The idea is that "for what" cannot become a reality until and unless the answer to the "to whom" question satisfies a broad-based coalition of citizen-participants. Failure here leads to the default option of no decision. Participants are also willing to support decisions that satisfy the "for what" component of accountability as defined by duly elected and constitutionally appropriate authorities; they are legitimate, case closed. But alliance participants do not want this to be the end of the discussion about the "to whom" question. Instead, they seek creative new methods to expand accountability to a broader array of interests (to whom) *without* harming the original intent (for what) contained in laws and regulations.

The ten outcomes complement and reinforce the broad-based, simultaneous accountability dynamic fostered by the Willapa Alliance's institutional structure, processes, norms, and management approach. Three cases are clearly located toward the "strong" end of the accountability continuum because each strengthens accountability to three of the four levels. Four of the ten outcomes, on the other hand, make accountability stronger at two levels, while concurrently preserving accountability to the other two levels. However, three of these four—the "Fisheries Management Plan," the "Chum Salmon Efforts," and the ShoreTrust Trading Group "Economic Development" outcome—strengthen accountability at the individual and community levels, while the fourth outcome—the "Watershed Restoration Partnership Program"—strengthens accountability to the community and state levels. The "Changing Worldviews" and "Taking Action on Private Land" outcomes also offer strong support for the idea of broad-based, simultaneous accountability, in that accountability is strengthened for the community and state levels, while simultaneously being weakened at the individual level.

However, while the outcomes as a whole display relatively strong accountability across levels, accountability to the national (macro) level is weaker than in the HFWC and Applegate Partnership cases, because only one outcome strengthens accountability to this level (compared to 5 out of 9 for the Partnership, and 4 out of 11 for the HFWC). Yet this does not mean that accountability to national concerns is weak in an absolute sense; seven of the ten outcomes preserve accountability to this level. Moreover, unlike the other two cases of GREM, wherein large parcels of land are federally owned, the accountability relationship with the national level is no doubt directly affected by the fact that only 3 percent of all land in the Willapa Basin is federal.

There is, of course, more to the accountability story than the matter of assessing outcomes against the individual, community, state, and national levels. Chapter 7 examines the lessons from the three cases of GREM as they apply to another level of accountability of central importance to the study of environmental policy: consideration for future generations. The analysis focuses on how the new governance arrangements help translate the goal of environmental sustainability into reality.

7

The Temporal-Environmental Dimension of Accountability: Building Community Capacity and Commitment to Sustainability

Traditional definitions of democratic accountability focus on politically derived authority relationships, obligations under the law, and the mechanisms designed to produce desired behavior or to prosecute indiscretions. The scales or levels of accountability are typically political and administrative in scope, and it is common to think of accountability in primarily jurisdictional or geographic terms. Yet the environmental policy field is a special case with respect to accountability, and no analysis is complete without giving due consideration to a nonstandard, temporal dimension of accountability. This means that accountability needs to be examined from the perspective of environmental sustainability, or future generations of citizens.

The starting point in any discussion about environmental sustainability is usually the famous 1987 Brundtland Commission report, *Our Common Future*. It defines environmental sustainability as "meeting the needs of the present without compromising the ability of future generations to meet their own needs" (The World Commission on Environment and Development 1987, 43). Many have taken this basic concept and attempted to flesh it out in fairly simple terms. Euston defines sustainability as "[a] condition in which social systems and natural systems thrive together indefinitely" (as taken from Hempel 1999, 47). Hempel (1992, 3) finds that it is "a process of creation, maintenance, and renewal that persists in balance with the process of decline, death, and decay." Others have taken the concept and broadened it to include specific reference to a complex array of community-based policy needs, including urban sprawl, education, and safe neighborhoods (e.g., President Clinton's Council on Sustainable Development, 1996). Yet, while most agree that there is little clarity as to what constitutes a proper definition of sustainability,[1] "as a practical matter, sustainability can mean any important changes in values, public policy, and public and private activity that moves communities and individuals toward realization of the key tenets of ecological integrity, social harmony, and political participation" (Mazmanian and Kraft 1999b, 18; Hempel 1996, 1998; Prugh, Costanza, and Daly 2000).[2]

The participants in the Willapa, Henry's Fork, and Applegate cases of grassroots ecosystem management explicitly recognize the environmental-temporal dimension of accountability; they aspire to sustainability via the sustainable-communities route.[3] In fact, it is almost as if the rhetoric and institutional logic of these efforts have taken a page out of the sustainable-community playbook, because both seek communities "in which economic vitality, ecological integrity, civic democracy, and social well-being are linked in complementary fashion, thereby fostering a high quality of life and a strong sense of reciprocal obligation among its members"(Hempel 1999, 48).[4] The approach to sustainability thus necessarily rejects the more traditional top-down, expert-led processes for deciphering and achieving sustainability[5] because it is believed that experts acting alone are inherently unable to deal with the complexity of the sustainability challenge and that long-term environmental policy success necessarily must involve the citizenry ultimately responsible for translating sustainability theory into on-the-ground results.[6] A sustainable community therefore must craft decision processes that actively and credibly engage affected citizens and stakeholders, the very people who must put the theory of sustainability into practice. As the Willapa Alliance (1998b, 160) notes: A sustainable community needs to be developed by the people who make up the community. It cannot be designed by a consultant. It cannot be implemented by experts hired specifically for the project. It needs to be implemented every day by the people who live and work in the community.[7] At heart, then, participants see their efforts as "a visionary project . . . [aimed at promoting a] paradigm shift in the way the whole community understands itself," both in the relationships among community members and the community's relationship with nature (Hollander 1995c, 8–9).[8]

Participants in these alternative governance arrangements are thus crafting and seeking to maintain institutions and practices in support of sustainability that, in essence, creates community anew, although the "new" communities in question are clearly works in progress. If done properly, the new community will possess additional capacity for achieving sustainability and a heightened sense of collective purpose centered on sustainable development. In the three cases studied here, the factors of chief importance to the added capacity for, and commitment to, sustainability are the

• Institutional framework—the mission, norms, processes, and management approach—used to manage the interface between humans and nature
• Creation of a unified, integrated community-based network
• Promotion of an informed, skilled, and engaged citizenry
• Incentivization of sustainability
• Explicit reliance on a broad knowledge base
• Development of sustainability indicators

Mission, Norms, Process, and Management

Through their missions, participant norms, decision processes, and management approaches, the Henry's Fork Watershed Council, the Applegate Partnership, and the Willapa Alliance are infusing their communities with the value of sustainability and providing a new, more appropriate institutional framework for managing the interface between humans and nature.

The crosscutting "environment, economy, and community" mission statement, by valuing each element equally, emphasizes interdependence by recognizing that the policy sectors are linked in a such way that long-term success (sustainability) in one arena is dependent on success in the other two. At the same time, the holistic ecosystem management approach recognizes that the watershed's "web of life" is interconnected such that individual, fragmented decisions by stakeholders in the watershed often affect the health and well-being of other resources, habitats, and people. Both cases define or frame environmental sustainability as inseparable from the sustainability of a healthy economy or a healthy community, thereby directing decision-making efforts to find the kinds of workable compromises that simultaneously promote the multiple policy goals.[9]

Participant norms reinforce the commitment to sustainability.[10] For example, participants are asked to envision themselves as embracing a dual role—that of community member and as representative of a particular interest (e.g., farmer, environmentalist, agency official). The "dual role" norm obligates participants to take a broader view of problems affecting the ecosystem or watershed, to take a Pogo-inspired "them is us" perspective, and to recognize the interconnectedness among people and policy sectors.

Taken together, the mission, management approach, and norms lead to a lack of tolerance for decisions that only benefit one particular segment of the community, or one element of the three-part mission at the expense of others. The incorporation of sustainability as a critical public value in the decision process also adds legitimacy to the goal of sustainability as well as to the practices necessary to achieve sustainability. The legitimacy attached to sustainability increases to the extent that a broad cross-section of interests and community leaders are represented. Increased legitimacy makes it easier for community residents to adopt sustainability practices on their own—for example, whether it is the conscious choice of "green" alternatives by industry interests, or a willingness on the part of civic and political leaders to honor the economic and ecological interdependence within and between communities (Hempel 1999).

Moreover, the institutional framework promotes sustainability because it organizes according to ecological scale (e.g., watershed) as opposed to political-administrative scale, adopts a proactive orientation that seeks to identify and prevent problems before they occur, and more successfully mimics the flexible, adaptive, and cyclic character of ecological systems (see Costanza and Folke 1996). The key aspect of the latter point is the adaptive management style. Adaptive management emphasizes learning through experimentation. Ideally, adaptive management increases responsiveness to ecosystem problems by promoting the values of continual innovation and adaptation in response to changing conditions, problems, and degree of success (or failure) enjoyed by solutions.

Creating a Unified, Integrated Community-Based Network

The institutional arrangements employed by the three cases are also helping to create what some have identified as an essential component of a place-based community—a dense set of networks that can be called on for communication, informal decision making, and action.[11] An environmentalist involved in the HFWC puts it this way:

The relationships with Dale Swensen and all of these other folks have resulted in the creation of networks within the community that simply did not exist before. When something comes up now, people are more prone to ask, "Well, who could help with that?" rather than arguing about jurisdiction and responsibility. It is not like the more traditional linear kind of thinking anymore, it's about networks. These relationships/networks create a new kind of problem-solving skill based on connecting community members together. The connectedness of the network creates a critical mass of people focused on problems common to the watershed. It creates new opportunities for passing on information and solving problems. That's what happened with a rancher who was struggling to maintain his water diversion structures on his land above Island Park Reservoir. He contacted the HFWC; he had heard about our other efforts in the watershed [e.g., the Sheridan Creek restoration project]. We knew right away that here was a rancher trying to do the right thing, fix his structures, conserve water, and, by extension, help the environment. But we also knew that he did not have the financial wherewithal to do it alone. We said, "Here's a community member who needs help and what a great opportunity to build another bridge to the ranching community." And it turned out that the Council, through an Idaho Fish and Game grant, could help. (personal interview, 1998)

The new dynamic connects the series of separate organizations and networks representing narrow, often self-contained segments of the population (e.g., irrigators or timber interests have one network, environmentalists another) to produce a more unified, integrated, community-based network whose primary purpose is a sustainable community. The transformation is analogous to the difference between the weakness and fragmentation evident in a shattered piece of glass lying on the ground and a multicolored, multishaped mosaic that has been welded together to form a stronger, more integrated whole. In short, the new network strengthens the capacity of the community to act collectively in several ways.

First, the process of working together helps diverse elements of the community, whether it is old-timers and newcomers, or commodity interests and environmentalists, construct a common community history. The common history becomes a way of defining the community by articulating the issues that matter. The important issues might be those that come up repeatedly, those that concern diverse community constituencies, those that pose potentially catastrophic consequences for significant numbers of community residents, and so on. By focusing on issues like changing patterns of land ownership, different uses of water resources, concerns over particular forest management treatments, or traffic congestion, for example, the common history provides a starting point for building new relationships and for constructing a shared vision of what the community wants to become (Priester 1994, 121; Hannum-Buffington 2000, 20–21).

Second, deliberating together and cooperating with others on efforts providing communitywide benefits breaks down negative stereotypes and leads to new, positive working relationships grounded in trust. An unaffiliated citizen-participant in the HFWC argues that this is "absolutely critical to building trust within the community. . . . To the extent that we investigate and cooperatively pursue projects that help all watershed residents gain something, trust will follow" (personal interview, 1999). An environmentalist member of the Applegate Partnership agrees: "Rather than being adversaries, we've come to realize that we have a great deal in common. We still disagree on lots of things, but the partnership helps us understand that we're also neighbors with a common stake in the health and well-being of our community" (personal interview, 1999). The newly minted trust encourages people to view others as partners and neighbors rather than as adversaries, and to communicate in a direct, open, and honest manner instead of hiding information for strategic advantage. According to a growing number of scholars, these developments help to unify the community, assist attempts to solve problems of importance to the community, and, more generally, increase community capacity for effective self-governance (Fukuyama 1995; Putnam 1993).

Third, the extensive networking means that institutions and decision makers who used to be inaccessible to many in the community, or were accessible, according to one environmentalist, only after "quite a bit of . . . moaning or legal action . . . are now only a phone call away because of the trust that networking has created" (personal interview, 1999). Citizens involved in the HFWC, including an environmentalist and a recreation interest, point to Harriman State Park as a prime example. When first approached about getting involved with the council and cooperating to manage resources that either were in the park or affected the park, park managers were "reluctant to jump in with both feet." Now, however, they are very enthusiastic about the council's collaborative format because they believe the council has helped them more effectively manage park resources, whether in the area of trumpeter swans, riparian restoration along the Henry's Fork, or simply taking care of upstream problems such that the park itself experiences fewer resource problems (personal interviews, 1998).

Fourth, the unified, integrated network strengthens the capacity for collective action by facilitating the creation of informal decision-making institutions to complement existing formalized arrangements. Two examples illustrate the point. Over in the Henry's Fork watershed an informal decision-making institution has developed to govern water releases from Island Park Reservoir. The Fremont-Madison Irrigation District (FMID) controls water releases and prioritizes them according to water rights claims by downstream irrigators. Yet, beginning in 1998, FMID has shown a willingness to be more flexible by releasing additional water to benefit the environment (e.g., to combat dangerously high water temperatures) at the request of Jan Brown, executive director of the HFF. There are limits to this arrangement—there must be "extra" water in the river. Thus FMID is unlikely to be very flexible during low-water years. But no one interviewed for this project can imagine the institutional change *without* the years of working together on the council and the creation of new relationships and trust among segments of the community who traditionally never had a reason to communicate with each other (except through lawyers), much less cooperate for the sake of the environment.

In the Applegate Valley, responses by federal agencies to the 1995 Salvage Logging Rider[12] are also instructive. The rider redefined salvage timber so broadly that there was no limit to the volume of healthy "green" trees included in a salvage sale.[13] Traditionally, no more than 15 to 25 percent of all trees in a particular salvage sale could be "green."[14] In addition, the rider expedited timber-sale procedures, exempted salvage sales from standard administrative review and appeals processes, and severely restricted judicial review for such sales.[15] Further, the salvage logging rider assumed that all sales automatically complied with all federal statutes governing timber agencies (e.g., National Environmental Policy Act, Endangered Species Act, National Forest Management Act, and so on) and "all other applicable Federal environmental and natural resource laws" (Gorte 1996, 2).

Many agency officials in USFS forests and BLM management districts took advantage of the enormous discretion offered by the rider to boost their timber-harvest levels, often by resurrecting "old" sales that had previously been denied for violating one or more environmental laws (Wilderness Society and National Audubon Society, 1996; General Accounting Office, 1997). Federal resource managers in the Applegate Valley, however, did not take advantage of the new opportunity to get more of the cut out, despite having experienced a sharp decline in timber-harvest volumes since the 1989 spotted owl court decision. Local agency officials "did not change the methods or timing of timber sale projects" (Rolle 1997b, 2), continued with selective cut, thinning sales already planned or in process (Shannon, Sturtevant, and Trask 1997, 19), and did not permit salvage sales located in areas that partnership participants had agreed were ecologically sensitive areas. In each case, the long-term cost of breaking the trust with community members was deemed too high to justify the short-term gain (personal interviews, 1998; Marston 1997b; Rolle 1997b).

Finally, the improved capacity of the community to act collectively, or to govern more effectively, does not occur in a substantive policy vacuum. Rather, it is governance effectiveness for a specific purpose. The unified, integrated, community-based network outcome is about cultivating a heightened sense of collective purpose in each place that is centered on the sustainability message inherent in the tripartite "environment, economy, and community" mission. It is evident in the comments of "core" members—those who regularly attend meetings and who exhibit the kind of passion and commitment to sustainability that drives the network forward. A federal official explains:

I was a skeptic at first. I didn't really believe in the partnership's mission. But as I listened, I became more open to hearing what the dream could be like and seeing the opportunities for . . . experiment[ing] with the whole sustainability issue, treating communities and forests and lands together, and working across diverse people. It really became for me an archetype of what we could do for the world. I mean it became a very, very powerful belief system for me personally, that if we can't pull it off here, we can't pull it off as a planet. And so I became . . . motivated by a genuine belief that this is absolutely the right thing to do for the Applegate . . . [and] that this will teach us lessons about sustainability for the future, about how we can solve [such] problems. (personal interview, 1999)

And an environmentalist adds:

The ecosystem is dynamic, it is constantly changing. . . . So what we're talking about is being flexible enough to cope with the changes . . . and so we talk about ecological sustainability. But we also need to talk about economic sustainability and community sustainability. . . . The fear is that we're now doing management on this land that over the years is not sustainable, which means that people—farmers, ranchers, loggers, our fathers—will be out of business based on the way that we have always done it. Yet people want to keep doing business the way they've always done it. . . . The reality is that we really need to reevaluate, to look at the ecological carrying capacity . . . the geologic, hydrologic carrying capacity of this watershed. Then just start to accommodate that. Ask what can the land support and work from there. . . . This will open our eyes to other opportunities and then we can start trying to change hearts and minds. (personal interview, 1999)

A vibrant, strong "sustainability" network thus has the capability to induce behavior motivated by enlightened self-interest—the understanding that what is beneficial for the collective community can also provide important long-term benefits to individuals. Self-interest is still treated as an important key for shaping behavior; this much is clear from the multifaceted emphasis on environment, community, *and economy*. The probability for success in a sustainable community is enhanced to the extent that economic activities (i.e., the profit motive) are aligned with the dynamic of ecological systems (see the section titled "Incentivizing Sustainability"). Yet participants accept that self-interest is malleable, that it is profoundly shaped by social interaction with others, and that a unified, integrated network focused on sustainability provides a constant reminder that individual self-interest should be weighed against the collective, long-term interests of the community, including future generations.

Over the long term, then, the work of the new network will be eased to the extent that it reaches into all corners of the community and elicits the same kind of passion for and commitment to sustainability from much larger numbers of community residents. Participants know this will not be easy. They seem to know, as expressed by one environmentalist-participant, that tackling sustainability successfully over time is likely to require

a massive community paradigm shift. . . . We are talking about pulling rather than pushing, about moving whole communities systematically toward sustainable economic development. That is pretty heady stuff because we are talking about changing the way people live and act and react to their ecosystem, and develop their economy and think about community health. That is extremely complex. . . . But . . . getting to sustainability is absolutely essential. If we don't do it, we'll lose the pristine character and quality of life that make our place so special. (personal interview, 1998; See also Hollander 1995c, 8–9; Marston 1997c, 1)

Improving Citizen Capacity for Promoting Sustainability

There is a growing consensus among public managers, scholars, and development officials that success in achieving sustainable development, or any other public goal, requires more than simply the endorsement and pursuit of the concept by government officials and natural resource agencies. Instead, success requires the active consent and support of ordinary citizens (Bell and Morse 1999; Box 1998; Chertow and Esty 1997; Johnson 1997; Knopman 1996; Scott 1998; Snow 1996; Steel and Weber 2001). Toward this end, the three cases of GREM are trying to increase community capacity for sustainability through the creation of a more informed, skilled, and engaged citizenry. The belief is that such a citizenry will have both a greater interest in promoting sustainability for the long term and a greater capability, thus increasing the probability that the interest in sustainability will be matched by results. The new governance arrangements do this by promoting a heightened awareness of what sustainable practices are, helping citizens recognize when something is amiss, and creating a cadre of citizens better able to monitor decisions affecting sustainability.

Education through Community Outreach

The Henry's Fork Watershed Council, the Applegate Partnership, and the Willapa Alliance use a variety of methods to engage and inform the community about the status and potential of natural resources in the watershed. The three efforts also educate residents as to how such resources are connected to economic health and how they can be sustained. There are two main categories of community outreach activities: active and passive. Active outreach involves purposive attempts to engage and inform community members, whether through the staging of events, meetings, and activities, or through the distribution of information, written or oral, pertinent to the mission. Passive outreach, on the other hand, involves making resources available for community use, but actual "education" is dependent on contact being initiated by community members themselves. Outreach efforts remind citizens to adopt a different mindset when it comes to natural resource management, namely, that management is the responsibility of everyone, rather than being the sole responsibility of public agencies.

Informing the Public about New Governance Activities   Active outreach is alive and well at the HFWC and appears to be increasing. The Council, recognizing limited participation by citizens in Teton County (the southern reaches of the watershed), is making a concerted effort to hold more meetings in Driggs, Idaho. Bimonthly meetings are advertised with individual mailings to over 200 people as well as through press releases to local newspapers. The notices mailed to the 200-plus people on the mailing list include the agenda for the next meeting as well, "so that everyone who plans to attend has a chance to mentally prepare for the meeting" (personal interview, 2000). A quarterly newsletter published by the Henry's Fork Foundation, a leading member of the HFWC, now reaches 1,800 people, up from 350 in 1992. Moreover, the newsletter has been redesigned to reach a general audience; the technical, scientific thrust that has been the hallmark of past newsletters is still there, but the information is now more accessible and is complemented with other human-interest stories about the watershed region.

Beyond public notices of meetings in the newspaper, the Applegate Partnership also sends meeting minutes and agendas out to the people on the partnership mailing list, including environmental groups, "so that they know exactly what we're talking about and planning to do" (personal interview, 1999).[16] Yet perhaps the most important public outreach effort by the partnership is *The Applegator*, a community newspaper published by the Partnership on a bimonthly basis. *The Applegator* is distributed free to all residents of the watershed and, if requested, to people outside the watershed. The paper tries to encourage residents to identify themselves with the Applegate community by adopting a nonpartisan, "us and them" approach for reporting community events and Partnership activities.[17] A citizen-member of the partnership explains that "it's set up so that you can't say this paper is Republican, Democrat, Libertarian or a tool of the timber industry, etc. . . . although every group will tell you it's the opposite. Timber folks will tell you, 'It's an environmental paper.' Environmentalists will say, 'Oh yeah, that's definitely an agency timber rag.' And that tells me we must be doing a good job because we're putting in a considerable amount of material from all these different groups . . . and people are obviously reading it. Where else would they be able to come up with their opinions?" (personal interview, 1999). Such was not always the case. In the beginning, observes a federal official, "you'd go to the Post Office and the trashbins were just full of them. . . . Now, you go anywhere in the Applegate, way up East Gulch Road or anywhere else, and everybody reads *The Applegator*. So it's become an important communication forum for the community" (personal interviews, 1999 and 2000).

The Willapa Alliance reaches out to residents with their biennial "sustainable indicators" summits and several "conflict resolution" forums each year. The public forums typically draw several hundred people to explore policy issues of importance to the community. A good example is the October 1997 forum on the relationship between endangered species, healthy habitat, natural resource users, and regulations.[18] In addition, local newspapers, primarily the *Chinook Observer* and the *Daily Astorian*, are an important outlet for such items as public notices of meetings and a regular Willapa Science column in the *Chinook Observer* (Willapa Alliance, 1995, 6). The Alliance also had both papers distribute over 9,000 copies of the executive summaries of the WISC Summits as inserts in order "to further engage the Willapa community-at-large" (Willapa Alliance, 1996e, 5). Further, the alliance maintains a membership list that numbered 9,300 as of March 1998. Many members live outside the area (personal interview, 1998; Willapa Alliance, 1996, 1997a).

In addition, all three governance efforts have published brochures highlighting the intricate linkages among the community, economy, land, and natural resources, more generally. In *A Home for All of Us,* the Applegate Partnership (1996a) and Applegate River Watershed Council seek "to provide [citizens] with an understanding of the interconnection of all varieties of life in this valley . . . [and] to offer simple guidelines to responsible practices for interacting with both the natural resources and your human neighbors. The goal is to increase awareness of the impact of our actions so that future generations can also enjoy and appreciate this very special land known as the Applegate Valley" (3). The pamphlet offers practical advice on such things as getting along with neighbors (4–5), reminders of how to "care for the land" (6–7), brief lessons on forest health (9), definitions of key terms associated with water resources (10), advice on "things you can do" to "help transform watersheds and streams into community treasures" (11), and more than three dozen phone numbers of government offices.

The HFWC has cosponsored a similar effort with the Teton County Economic Development Council (1997) called *Welcome Home: A Homeowner's Handbook for Living in the Teton Valley.* Because "everything that makes up our lifestyle has a consequence to the natural world," the sixty-page book is designed to encourage kinds of "informed choices" necessary for ensuring that the area "continues [as] a natural paradise into the future" (i). A broad range of topics are covered, ranging from the importance of healthy habitat for wildlife populations, the role of farming, the spread of noxious, nonnative weeds, advice on grazing practices and responsible horse ownership, and recycling, to how to get involved in the community, among other things.

The Willapa Alliance, on the other hand, worked with the Washington State chapter of the Nature Conservancy to publish a book on the Willapa Basin. *A Tidewater Place: A Portrait of the Willapa Ecosystem* (Wolf 1993) "became a best seller in the Willapa area, I think, because it highlights the wonderful, magical qualities of the Willapa. It made people proud and it helped them understand better why their place is special. And this is why the book is one of the best things we did. . . . [Because] if they understand it is a special place and why it is special, it becomes a way to get people to pay more attention to it as well as the connections between the community and the natural resources of their special place. The hope is that they will then take more care of it" (personal interview, 1998; also see Hollander 1995g).

Getting Involved in Community-Based Organizations and Activities
The actors in GREM have also engaged in a number of attempts to get the message of "environment, economy, and community" out through presentations in local elementary schools and through increased interaction in community-based organizations and activities. A staffer from the Henry's Fork Foundation, the organizer of the school efforts for the HFWC, explains that because "the schools have welcomed us with open arms . . . we've taken to the schools more than in the past. This is a long-term investment on our part, but we like it because the kids don't have any preconceived notions about, well, this is just an environmental group coming to shove something down my throat. They are very open to the message of conservation and restoration of natural resources" (personal interview, 1999). School presentations in the Henry's Fork area are no

longer limited to the Ashton, Idaho, area schools; they are now made throughout the watershed and occur, on average, once every two months, primarily in grades 5 and 6. The efforts have not only spurred "a lot more inquiries" on the part of teachers as to how the sustainability message can be integrated into their curriculums but have led to "some really good feedback from the teachers as to how we could make the [materials] better." The interest in the presentations has also "created more interest and inquiries from other grade levels [i.e., other than grades 5 and 6]" (personal interview, 1999).

HFWC participants are trying to insert themselves into a variety of more general community-based activities as well. They see this as a good way to "help the community reach some of their goals" and to increase the chance that others will "perceive [us] as interested in the economy, the ecology, . . . [and] the community, a perspective that treats the whole of the watershed" (personal interview, 1999). Several council participants have also started to attend local Chamber of Commerce meetings on a regular basis and are chairing a committee to make sure that the annual River Regatta—a race down the Henry's Fork—happens. The regatta "brings commerce here and we are interested in bringing commerce here, but it's also our way of managing the impact from the inside looking out. By actively managing the regatta now [and in the future] . . . if the need ever arises, we are in a much better position to speak for . . . the environment in case the regatta starts to get too big and starts to have negative impacts along the river" (personal interview, 1999). Moreover, the River Regatta is a vehicle for raising money to replace a county-owned fishing bridge and handicap access to the river that were destroyed by an ice jam in the winter of 1999.

The Willapa Alliance has focused its efforts on what it calls place-based community education. There are four main components: a school resource guide, The Nature of Home Program, Student Institutes, and primary sponsorship of the Willapa Week community celebration each spring. In cooperation with local educators, the alliance has crafted a 172-page school resource guide for grades K through 12. The guide, titled *Places, Faces and Systems of Home: Connections to Local Learning,* "is a toolbox to help [educators] learn and teach about Willapa by utilizing local resources. It is our hope that this Resource Guide . . . will encourage and support . . . learning opportunities for all community members based on our home system, The Willapa, [in order] to promote a collective shift towards sustainable development" (Willapa Alliance, 1998a). The resource guide contains over 140 listings of Willapa-specific resources, background information on the area, concepts critical to community sustainability, and a computer diskette with a searchable database of over 300 resource listings likely to be of use to educators (Willapa Alliance, 1998b).

The theme for the Willapa Alliance's *The Nature of Home* (NOH) *Program* is "In Willapa, life is wet, and water is the ultimate currency." Water is currency because it brings the community together; "water 'funds' Willapa's environment, economy, society, and culture" (Willapa Alliance, 1998a). A local community education specialist in cooperation with a local advisory task force directs the NOH program. It involves the utilization of educational and scientific tools developed by the alliance, such as the

- *Willapa Indicators for a Sustainable Community* (WISC) report
- Willapa Geographic Information System (WGIS)[19]
- Understanding Willapa Educational CD
- Willapa Fisheries Recovery Strategy (WFRS)

The NOH program "strives to inform the present and future adult generations of their host ecosystem, to prepare them to be civic participants and leaders, and to aid the evolution towards a stewardship-minded employers and employees. Most importantly, we utilize specific, local field sites . . . to make a connection that is not only intellectual, but personal as well, to foster one's internal stewardship values" (Willapa Alliance, 1997b, 1). A variety of subjects are taught, all centered on the intimate, essential connections among economy, ecology, and community. Subjects include, but are not limited to, sustainable development, biodiversity and rural watershed communities, Willapa geology, and ecological processes (Willapa Alliance, 1997b).

Further, there are the annual weeklong Student Institutes, utilized by more than 100 area high school students to date. Student Institutes create opportunities for localized learning about sustainability through field trips, seminars with local experts from government agencies, schools, and the private sector, and projects wherein students themselves design and build hands-on, interactive, interpretive teaching tools. The lessons learned are then displayed at the Spring *Willapa Week—A Celebration of Home,* a more general community gathering designed to help residents learn more about the sustainability potential of the Willapa (Willapa Alliance, 1998a).

The Applegate Partnership encourages similar efforts in local schools, although not as extensively as the Willapa Alliance. Yet it is far more active in reaching out to other community groups than the other two governance efforts: "Something that we're particularly interested in . . . is building capacity for sustainability over time, first with the individuals involved in the partnership, and then those individuals can network out to work with other community groups, whether they are focused on health care, or friends of the library, or the Apple Core looking at economic opportunities, or land use planning efforts" (personal interview, 1999). Productivity is a high priority: "We have always said that the partnership cares more about whether the job gets done than who gets the job done" (personal interviews, 1999). In this respect, the partnership is less about competition for turf or credit than about achieving results on the ground.

Professional and Political Outreach   A final category of active outreach is professional and political outreach. This captures the willingness of participants to talk at length with people from outside the community about their accomplishments and the basics of their community-oriented approach to understanding, restoring, and protecting their special place. As part of this effort, participants attend watershed and natural resource management conferences around the country.[20] They also testify before Congress and state legislatures. For the HFWC, cofacilitators Jan Brown and Dale Swenson report on the council's progress to the Idaho state legislature once every year as required by law. Dan'l Markham, executive director of the Willapa Alliance, and Jack Shipley of the Applegate Partnership, among others, have each testified before their respective state legislatures a number of times and before the U.S. Congress at hearings on Community-Based Approaches to Conflict Resolution in Public Land Management (U.S. Senate, 1997).

Participants also make it a point to nurture relationships with political and bureaucratic leaders, whether it is local county officials, federal agency heads like Bruce Babbitt of the Department of the Interior, or

James Lyons, Undersecretary for the Department of Agriculture, or influential elected officials like Governor John Kitzhaber (D-OR), U.S. Senator Mike Crapo (R-ID), and Senate Majority Leader Sid Snyder (D-WA) (personal interviews, 1998–2000). For Brown of the HFWC, a case in point is her relationship with the Idaho Director of Parks and Recreation, which resulted in an invitation to address the state-appointed Parks and Recreation Board when it met in the area: "Often I find that state-level organizations like the Parks and Recreation Board don't know anything about our efforts other than at a superficial level. . . . They get little snippets of information here and there. But by my being there, making the time to talk to them . . . it exposes them to the total approach and what we're trying to do and how we're trying to do it. . . . Normally I just try to build up their enthusiasm for [the collaborative, community-based] watershed approach, and I introduce the HFWC as a way of improving both communication and performance" (personal interview, 1999). Brown is also a familiar lobbying force on Capitol Hill in Washington, D.C., making several trips each year to inform and maintain relationships with elected representatives. Over in the Applegate, Jack Shipley "often hobnobs with political heavyweights. He has contacts at the state level in Oregon and the national level" (personal interviews, 1999 and 2000).

Some of the political outreach efforts are attempts to engage regional and national environmental advocacy organizations. A citizen involved with the Applegate Partnership explains that participants met with leading national environmentalists on their own turf in Washington, D.C.,

to explain what [the Partnership] is trying to accomplish . . . [and] to say, "Please come to the table. We recognize that you have strong values and certain principles that you would like to see happening in terms of restoration and in terms of wilderness protection or whatever." And we invited the whole crowd to come, American Forests, the Sierra Club, the Wilderness Society, the World Wildlife Fund. . . . At different times, we have also called and encouraged the ONRC [Oregon Natural Resource Council] to come down and see what we're doing . . . come out on these field trips and see because it must be difficult to judge this stuff from afar. . . . [Only a few of the national environmentalists ever came to any meetings, while] the ONRC never came. (personal interviews, 1999)[21]

Passive Outreach   Passive outreach is also alive and well in all three efforts. As a result, anyone who has a concern about governance activities, or perhaps simply wants to know more about the state of the community or watershed, has several means for doing so.

The new governance arrangements have played a central role in coordinating and making more accessible various information databases pertinent to each place. The Applegate Partnership has been instrumental in coordinating the integration of BLM, USFS, and private-sector geographic information system (GIS) databases. The Willapa Alliance has done the same, creating a GIS record of the Willapa Basin in cooperation with Interrain Pacific, a private-sector GIS specialist. The HFWC archives a GIS databank of the watershed that anyone can access in the Ashton, Idaho, offices of the Henry's Fork Foundation (HFF). Moreover, the HFWC, through the HFF, takes the data-collection idea a step further by maintaining a small library. The combination of the library and the GIS databank is attracting additional interest and visits from academics and government (agency) officials at all three levels of government (local, state, federal). The additional visits are ascribed to the quality and scope (watershed) of the information available, and because, in some cases, the library and GIS data offer additional resources for double-checking and confirming the veracity of agency information (personal interviews, 1998).

The Applegate Partnership and Willapa Alliance have also been aggressive in researching and cataloging other community-specific information that is now part of the public record. The Willapa Alliance makes available to the public the WISC report, the Willapa Fisheries Recovery Strategy documents, and a more general database on Willapa's history and contemporary character (the Willapa Database), among other things. The Partnership, as discussed in chapter 4, has undertaken several studies that explore the Applegate's social, demographic, economic, and environmental characteristics as well as residents' views on a variety of different public policy issues. They have also commissioned an "Outreach and Education" project titled *Stewardship in the Applegate Valley: Issues and Opportunities in Watershed Restoration* (Applegate Partnership, 1995). The studies are on file at three different local libraries as well as in USFS and BLM offices (Applegate Partnership, 1996a). The Willapa Alliance follows a similar strategy for dissemination of information about the community. For example, the WISC reports and the Willapa Fisheries Recovery Strategy documents are archived at public libraries in South Bend, Naselle, and Ilwaco, Washington, the Pacific Museum in South Bend, and the Ilwaco Heritage Museum (Willapa Alliance, 1996a).

All three groups take meeting minutes that are available to the public. The most extensive of these efforts is in the Applegate case, primarily because the partnership has held so many meetings. Meeting minutes are archived in the USFS Star Ranger Station on Upper Applegate Road and in the public library, on the Applegate Partnership letterhead. The primary recorder is Phyllis Hughes, a member of the local Sierra Club chapter and a participant in 80 percent of all Partnership meetings.[22] Participants are grateful to have such consistency, as a federal official explains: "We're very fortunate to have an excellent recorder of what goes on in meetings. Through the recording you can follow our commitment to [the Partnership's] mission. You also know who committed to do what and whether it ever got done" (personal interview, 1999).

Skills, Expectations, and Sustainable Practices

Participants are convinced that the institutional dynamic improves citizens' skills and changes expectations in matters related both to governing and to sustainable practices. Through this process, citizens encounter new opportunities for empowerment, for building the citizenship skills critical to self-government, for accepting greater responsibility in governance, and for exercising local oversight and implementation.

The collaborative, deliberative, participative elements increase opportunities for citizens to engage in the primary political art of deliberating, or reasoning together. The active involvement can reconnect citizens to government in a positive way by giving them a stake in governing and help tame selfish passions through deliberation, information sharing, and a better understanding of the "big" policy picture affecting the community (Landy 1993; Sandel 1996, 5–6). As part of this, the new governance arrangements remind citizens of the need for a new attitude toward the responsibilities of citizenship. The new attitude recognizes and accepts the connections between civic responsibility, active engagement in public life, and a community's quality of life. This sentiment is captured by a citizen-scientist: "Instead of sitting around and waiting for government to act, to solve the contentious issues tearing our community apart, we're here to tell people that sometimes local people can take care of their own problems, and their own facilities [e.g., local parks. . . . We don't necessarily need government to do all of these things for us. It is our community. It is our responsibility" (personal interview, 1999). According to another participant, an environmentalist, it is also a matter of "convincing citizens to not let communities of interest from outside the community define the goals and aspirations of our community in absentia, by default" (personal interview, 1999).

Field trips to inspect various facets of the watershed give citizens first-hand experience with what an (un)healthy forest, riparian area, or tall-grass meadow, for example, looks like and the relationship between decisions and watershed health (sustainability). Regular presentations at meetings by scientists, agency based and otherwise, are designed to create a working familiarity with issues of importance to the area in question. Over the years, for instance, HFWC participants have had the opportunity to observe presentations on, among other things, trumpeter swans, rainbow and cutthroat trout, grizzly bears and their habitat needs, and the relationship between ground and surface water flows in the surrounding area. Willapa Alliance participants have listened to presentations on noxious weeds (e.g., spartina), fisheries health, the relationship between water quality and the productivity of local resources (e.g., oysters, cranberries), the differences in timber-extraction practices, and so on. The Applegate Partnership has witnessed experts discussing anadromous fish and their habitat needs, the historical status and changing character of local forests' flora and fauna, the tremendous diversity of the surrounding Klamath bioregion, and the connection between indicators and sustainable forest management, among other things.

The Willapa Alliance has also enlisted community volunteers in the fight for sustainability. Through its Salmonwalk Training Workshops, 150 local citizens, including economically displaced fishers, have been trained and certified as stream surveyors for habitat and salmon populations. The streamside survey information is utilized in the GIS system and the Willapa Fisheries Recovery Strategy. Similarly, the alliance has initiated local stream stewardship efforts modeled on the Adopt-A-Stream concept. Stream stewardship (e.g., the Bear River Enhancement Association for Resources and Salmon) involves training volunteers in stream surveying methods, and bringing a diverse group of watershed stakeholders—government, landowners, fishers, and timber companies—together through classroom and field education as well as field projects.[23]

In sum, the stake in governance creates ownership (i.e., added incentives to care about how a place is managed), while the field trips and additional knowledge empower citizens by making them more capable of understanding the impact of decisions on their place. Taken together, these elements increase the likelihood that there will be a cadre of citizens better able to monitor not only GREM decisions, but also the effects of federal and state agency decisions on sustainability. When combined with the extensive communication among community residents facilitated by public outreach efforts and the unified, integrated networks, the effects are even more powerful according to a federal official: "Instead of having just a handful of individuals interested in the design and implementation of a particular agency project, you've got hundreds of people now, that . . . are suddenly watching government actions" to see if promises are matched by performance (personal interview, 1999).

Incentivizing Sustainability

To achieve [our] goals, . . . [we] need to help others achieve their goals. The old way is a win-lose conflict, more for me, less for you. The new way is cooperation, a win-win deal, where there is more for everyone. (Dan Daggett, keynote address to 1995 HFWC Annual State of the Watershed Conference.)[24]

As practiced in the Henry's Fork, Applegate, and Willapa areas, GREM rejects the traditional coercive approach to natural resources management that "tend[s] to alienate the very people who can make good conservation happen—or who can block it through inaction, a never-ending search for loopholes, or just plain recalcitrance," and that "so few westerners—especially those living on the land—are apt to sit still for" (Snow 1997, 198). A classic case of resistance is the "shoot, shovel, and shut up" response by private landowners to potential Endangered Species Act listings. Others are afraid, hence unwilling, to improve habitat on their land because doing so may attract threatened species, thereby restricting their management options (personal interviews, 1998). The responses confound the intent of the law by contributing to a perverse outcome—a hastening of the demise of endangered species, either directly or indirectly through fewer acres of prime habitat (Mann and Plummer 1995; personal interviews, 1998 and 1999).

Instead, the new governance arrangements employ a cooperative, participative format not only because it is seen as capable of benefiting nature and the community as a whole, but because participants see it as better able to provide private benefits for individuals as part of the same bargain (personal interviews 1998 and 1999). Daggett (1995) aptly documents the logic behind many win-win deals in the "stories of ten ranchers who invited their neighbors, 'experts,' environmentalists, and others to work with them to find better ways to manage their rangeland. The [ranchers] speak with pride of revegetated lands, larger and more diverse wildlife populations—*and higher profits*" (Getches 1995, viii; emphasis added). The potential for gaining individualized benefits through the new governance arrangements incentivizes participation and consensus agreement on decisions. From the perspective of participants, the prospect of individualized benefits is also crucial for convincing at least some, perhaps the majority of, *private* landowners to voluntarily adopt and support different, more environmentally beneficial land, water, timber, mining, and livestock management practices. Incentivization thus becomes another bridge to the private sector because to the extent that a series of initial bridges are built and management success ensues, the probability increases that more landowners will follow in their footsteps. In any case, whether one bridge is built or many, the belief is that incentivization brings more private land into the management mix and increases the capacity to manage ecosystems as integrated, sustainable wholes.[25]

In general, participants agree that convincing private landowners to cooperate for the sake of sustainability does not always require higher profits, although the high profits–more sustainable practices combination is the optimal outcome. At a minimum there must be enough certainty that cooperation and the adoption of more sustainable practices will not result in a loss of income, whether through higher operating costs accompanied by the same level of production, or the same operating costs accompanied by a lower level of production. In other cases, it is matter of autonomy, the right to control what happens on the land, or, at the very least, a guarantee of choice (versus diktat) from an expanded menu of more sustainable alternatives (personal interviews, 1998).

The three cases of GREM are rife with examples of how the cooperative approach incentivizes participation and the embrace of more sustainable practices by private landowners. For example, efforts to identify and restore Yellowstone cutthroat trout populations in the Henry's Fork region, or to restore steelhead and salmon habitat in the Applegate Valley and Willapa Basin, require access to private lands for surveying and habitat assessment purposes. Yet, as an environmentalist makes clear, many landowners would rather not allow government officials onto their lands for "fear that the science and information might lead to more regulation, which usually means a loss of income. . . . We had an initial meeting with local landowners [in the Willapa], primarily tree farmers, egg farmers, and cattle owners, to gain their cooperation for our stream surveys for salmon. And the owner of a small wood lot, a tree-farmer, a smart guy who manages on a sustainable basis, got up and spoke very passionately. He said, I have a couple of miles of excellent fish habitat on my property, and for years a county road culvert has restricted fish breeding in that stretch of stream. I want to open that habitat up for the sake of the fish, but I'm concerned that if I do, I'll pay too high of a price because I'll lose the ability to manage the area around the water" (personal interview, 1998).

Through their actions over time, however, all three efforts have established reputations as organizations that develop and share information with landowners for the purpose of helping them manage their property/resources in a more efficient and environmentally sensitive manner (e.g., GIS databases; stream surveys on fish populations and habitat) (personal interviews, 1998 and 1999). The reputation as a collaborator, or partner, as opposed to a regulatory adversary who imposes added costs with no regard for the effects on economic viability, has "helped overcome lingering doubts about our mission and increasingly earned the respect and trust of [Willapa] basin landowners" (personal interview, 1998). The Willapa landowner concerned about letting Alliance members survey his streams ended up allowing surveyors access to his land and is now an integral part of the Willapa Alliance working to promote sustainability practices throughout the basin.[26]

Another example involves livestock management practices on the Diamond D Ranch in the Henry's Fork watershed. To the detriment of Targhee Creek's riparian ecosystem and water quality, the ranch had always allowed cattle to graze at will and to access the creek for watering purposes. However, in cooperation with the council and in exchange for a partial subsidy totaling $10,000, the ranch made significant changes to cattle management practices. The rancher agreed to fence off key parts of the stream, install a watering trough with a float (to match water supply with actual demand), and adopt a different, more environmentally benign grazing practice ("hub" grazing focused on the water troughs rather than the stream).[27] The net result was a more efficient usage of range grasses, a 70 percent savings in water usage, a healthier riparian ecosystem, and cleaner water (personal interviews, 1998 and 1999).

There also is the case of the stockyard business in the middle of the Applegate watershed, discussed extensively in chapter 4. As an alternative to shutting down an established economic component of the community, the owner received a one-time subsidy to help restructure his cattle operations, volunteer help to protect and restore riparian habitat, and information about how different management practices impact environmental quality. The cooperative approach maintained the economic viability of the business while ensuring progress toward greater environmental sustainability.

Others find that participation exposes them to new information regarding the interdependency between ecological processes and efficient business practices. This is what happened to some farmers in the Applegate Valley. They learned to recognize the interdependency between bats, a night-flying insect control essential to crop health and tree snags that often are home to bats. Prior to this, many farmers viewed tree snags as an expendable nuisance that got in the way of farming (personal interview, 1999).

The Willapa Alliance, on the other hand, has nurtured a "strong relationship" with the Shorebank Enterprise Group (formerly ShoreTrust Trading Group) based in Ilwaco, Washington, at the southern edge of the Willapa Basin. Shorebank's goal is to create new markets and market opportunities for quality, "green" products from the Willapa seafood, agricultural, and specialty forest-product sectors. The idea, according to Mike Dickerson, Shorebank's marketing director, is "to create a basic economic self-interest in protecting resources. [After all,] why would a business change its practices, and do things that may cost it more, if there's not an economic return" (as quoted in Johnson 1997, 32).

Further, the Applegate Partnership has played an instrumental role in bringing certified organic farmers together to share best practices and to market produce directly to consumers in local markets. It also encourages the development of specialty wood products from surplus timber—small-diameter trees, non-commercial-grade wood, and brush remnants—much of which is already headed to the dump or the slash pile (i.e., burned because of limited commercial value) (see chapter 4). Both cases promote community sustainability by increasing the economic viability of local businesses, creating new jobs, and otherwise helping people make higher profits. Of equal importance, the increases in viability, jobs, and profits increase the opportunities and incentives for residents to make their living by engaging in business practices that are more environmentally sustainable.

Tapping a Broad Base of Knowledge

GREM expands the concept of expertise beyond scientific, bureaucratic, and organized-interest expertise, and seeks to engage and catalyze available community assets.[28] Participants, including government agency representatives, expect that bringing "new," qualitatively different knowledge to the table will improve the effectiveness of the new governance regimes. Consequently, the likelihood improves that more of the primary mission, focused as it is on sustainability, will be achieved. First, federal and state officials find GREM to be a "refreshing" forum for new ideas and potential solutions that typically do not see the light of day (personal interviews, 1998 and 1999). Second, the broad knowledge base tends to increase the probability of a more comprehensive understanding of problems, possible solutions, and consequences. Third, participants believe that the reliance on a broad knowledge base does a better job of incorporating the essential features of the real, functioning social, political, and ecological orders of the community in question. Put differently, the new information increases the likelihood that the dynamics of human institutions—social, political, economic, and administrative—will be matched with those of ecological processes to the greatest extent possible.[29] Taken together with the ongoing, deliberative discussion format, the broad knowledge base tends to produce a more robust set of alternatives for solving problems, and a more reliable and realistic estimate of the parameters affecting program success.

The reliance on a broad knowledge base comes in part because participants accept that real-world problems typically do not fit neatly into the singular domains of traditional scientific disciplines, nor are they amenable to analysis excluding social impacts. From this perspective, the natural and hard sciences are critical for resolving natural resource problems, yet the science must still be applied in human societies. Thus social science is valued along with the physical/natural sciences (e.g., silviculture, biology, ecology, chemistry) and technical professional advice (e.g., engineering). In short, while technical expertise and "hard" science are important, they simply are not sufficient for understanding, much less solving, the sustainability conundrum, especially when it comes to a more sophisticated understanding of the self-organizing human systems that are essential to progress in the battle for environmental sustainability (Kates et al. 2001, 641; Scott 1998).

At the same time, there is concern that professionals suffer from what the public administration literature calls the trained incapacity problem. Bureaucratic experts, whether engineers, economists, foresters, and so on, are taught to analyze certain types of situations in a specific way, using certain assumptions, procedures, and decision rules. They also tend to work from a restricted menu of possible solutions. For example, if given a water resource problem such as flood prevention, engineers will likely settle on technical, structural solutions—a dam to capture and control the flow of water, or a series of levees built high enough to prevent flooding. The range of possible solutions centers on the technical mastery of the river rather than accommodating human settlement to natural flow patterns or employing nontechnical solutions to the problem (e.g., adjusting insurance rates upward to reflect the true risks and costs of flooding). High levels of trained incapacity increase the difficulty of learning a new set of premises and assumptions about how the world works. By extension, an expert suffering from trained incapacity will also have greater difficulty adapting solutions to fit changing circumstances, whether ecological, economic, or social in nature (Knott and Miller 1987, 172–181).

Moreover, the three cases of GREM are designed to include the input of citizen-generalists with a "community" perspective (see Scott 1998, 309–341; Goldstein 1999). This means that there is an explicit reliance on community-based "folk knowledge"—the individual and collective expertise of the community members most practiced or most familiar with a particular problem and the capacities of the ecosystem in question. Examples of folk knowledge include the history of watershed drainage patterns, the resilience of and changes in particular forest ecosystems over time, recollections of conditions promoting the health of riparian areas and fisheries, and stored memories regarding what works and what does not when it comes to interacting with nature.

It is as if GREM participants are putting into practice several of the chief lessons from James Scott's (1998) book *Seeing Like a State*. Scott describes the ideology informing traditional bureaucratic practice as well as "schemes to improve the human condition" as "high-modernist." High-modernist ideology is "a strong, . . . muscle-bound version of the self-confidence about scientific and technical progress, the expansion of production, the growing satisfaction of human needs, the mastery of nature, and, above all, the rational design of social order commensurate with the scientific understanding of natural laws" (4). The problem is that the high-modernist framework often feeds into "an imperial or hegemonic planning mentality . . . [that] is necessarily schematic; it always ignores essential features of any real, functioning social order. . . . [and] excludes the necessary role of local knowledge and know-how. [In addition to such] . . . practical knowledge, informal processes and improvisation in the face of unpredictability . . . [are] indispensable" to the policy process given "the resilience of both social and natural diversity and . . . the limits, in principle, of what we are likely to know about complex, functioning order" (6, 7). As such, the "formal schemes of order" favored by bureaucratic experts, grounded as they are in scientific management and imposed from above, *need* practical, local knowledge; they "are untenable without some elements of the practical knowledge that they tend to miss" and in many cases lead to tragic consequences, sometimes of epic proportion (e.g., Soviet collectivization of agriculture; Tanzanian "villagization") (Scott 1998, 7).

Two examples of how local and/or nontechnical, practical knowledge can help lead the way toward more sustainable practices involve the design and placement of concrete and steel headgates for managing water flows in the Applegate and the Henry's Fork area. In the Henry's Fork watershed, a flyfisherman with twenty years experience reading the direction and speed of water flows on dozens of rivers pointed out how the placement of a headgate was "out of whack relative to the natural direction of the stream flow. In high water the stream will flow around the edge, causing heavy sediment flows [detrimental to fish and other aquatic species] and eventually rendering [the structure] useless. It needs to be shifted in this direction and made wider" (personal interview, 1999). Less than two years after the headgate was built, and as predicted by the fly-fisherman, nature was starting to get the best of the engineer's design. Similarly, when a new head gate was installed on the lower part of the Little Thompson River in the Applegate, "the engineers decided to put a little wingwall on it that they thought would make it work better. [However,] the ranchers looked at it and said, 'You can't put that in, it's going to cause a serious erosion problem.' The engineers said, 'No it won't.' Well, of course, come the first high water, what the ranchers said was going to happen happened. So it had to be cut out" (personal interview, 1999).

Indicators for Measuring Progress toward Sustainability

The common wisdom says that sustainability indicators are a necessary tool for solving the sustainable-communities puzzle.[30] They provide the kind of feedback required for monitoring progress toward a predetermined, agreed-on "sustainability" scale. The belief is that the richer, more accurate database improves the probability of sustainability by helping communities to not only make better decisions in the first place, but to fine-tune and adapt programs over time as conditions change. Yet, precisely because the Applegate, Willapa, and Henry's Fork governance arrangements focus on sustainable *communities,* indicators focus on more than just ecological sustainability. Instead, the emphasis is on integrating a set of standardized measures "of economic, social, and ecological health that are designed to gauge a community's systemic balance and resilience over long periods of time" (Hempel 1999, 63).

Moreover, Heinen (1994, 23) points out, and many agree,[31] that "sustainability must be made operational in each specific context (e.g., forestry, agriculture), at scales relevant for its achievement, and appropriate methods must be designed for its long-term measurement." The questions are how and by whom? The conventional practice has been to let bureaucratic experts and scientists determine the specifics of sustainability indicators from the top down. However, the case studies illustrate a different approach to these questions.

Participants in the Henry's Fork, Applegate, and Willapa cases acknowledge that operationalizing sustainability in their specific place and choosing appropriate measuring protocols are political questions as much as they are scientific because they presuppose a major shift in the behavior and practices of community residents. In addition, participants see few, if any, clear-cut answers to the technical dilemmas posed by the search for sustainability, whether it is a matter of appropriate monitoring protocols, for example, or the criteria (indicators) used for measuring sustainability. Given this context, GREM participants prefer a decision process that combines a bottom-up decision process involving as many parts of the community as possible with the top-down input from natural resource agencies' scientists and administrators (personal interviews, 1998 and 1999).

Professing allegiance to the value of sustainability indicators (SI) and changing the process by which they are determined, of course, are not the same things as actually constructing a viable SI framework. At present, only one of the three cases—the Willapa Alliance—has made much headway in defining sustainability, selecting indicators, and crafting applicable monitoring protocols. Yet Alliance participants are the first to admit that their efforts fall far short of being definitive (Willapa Alliance, 1996g). Another case—the Applegate Partnership—clearly realizes the critical importance of SI to their overall efforts to achieve sustainability. To date, however, the partnership has taken only a few tentative steps in the general direction of a workable indicators system despite having first discussed proposed criteria and indicators for measuring sustainable forest management in September 1993 (RIEE presentation, September 10, 1993). The third case—the Henry's Fork Watershed Council—has not taken any steps in this direction.

Conclusion

New governance arrangements known as grassroots ecosystem management are trying to craft sets of institutional rules, processes, and management practices aimed at fostering sustainable communities. Fully connecting the dots between institutions and sustainability requires an examination of both formal and informal institutions, and of the roles played by education, incentives, different kinds of knowledge, and sustainability indicators. Connecting the dots also requires recognition that success ultimately demands a heightened level of commitment to a sustainable community from those who live on the land and whose liveli-hoods have the greatest impact on the land. The efforts thus focus much of their energy on strengthening the bonds of community through pro-cesses of shared decision making, the encouragement of new relation-ships, and the facilitation of new cooperative networks (Shipley 1996; Johnson 1997; Sturtevant and Lange 1995).

Clearly, creating a new community committed to sustainability is not a task for the faint of heart. It is complex. It is messy. It will engender at least some resistance from residents seeking to preserve the status quo. And it is likely to take a generation or more.

Nor will participants have an easy time convincing hard-core contem-porary environmentalists that GREM communities are actually on track toward a sustainable future. This is because the new governance arrange-ments are fundamentally about sustainable communities—humans are included in the equation and existing land-use practices, extractive or otherwise, are treated as legitimate although in need of serious improve-ment in terms of their environmental impact. The sustainable-community approach is thus about balance among several goals and the competing values of a diverse society. It is pragmatic in recognizing that humans are an integral, legitimate part of ecosystems and therefore deserve to coexist with nature. And ultimately, it is not revolutionary because incremen-tal progress in the general direction of environmental sustainability is acceptable.

The sustainable-community, humans-with-nature approach thus is at odds with a purist, or "deep-green/ecology" perspective on sustainability. Deep ecologists conceptualize the same relationship as nature first, hu-mans second, and are perfectly willing to coercively restructure human choices in order to ensure success (as defined by them) (Bahro 1986; Naess 1983). The deep-green approach, for example, would most likely challenge the location of many rural communities throughout the West-ern United States as incompatible with the carrying capacity of the sur-rounding ecosystems, and therefore as incapable of sustainability *in ecological terms* and as prime candidates for relocation (Brower 1995). From this perspective, figuring out how the new institutional arrange-ments facilitate sustainability misses the point because the inquiry fails to ask a more fundamental question—should humans have settled in a particular area in the first place?

Others might argue that the institutional framework for ensuring sustainability is incomplete, especially in the case of the Henry's Fork Watershed Council, because the attempt at sustainability is ad hoc—it occurs on a project-by-project basis. And although each decision is subjected to scrutiny in terms of its relationship to ecosystem health and sustainability, there is only a nascent infrastructure of sustainability indicators for measuring and monitoring outcomes against the desired "sustainability" outcome. Without such measures, it is difficult to assess progress toward sustainability in a comprehensive fashion.

Nonetheless, if sustainable communities are about humans living together with nature, it is plausible to accept that these communities are making progress toward sustainability in a number of ways. The ecosystem management approach is designed to mimic and adapt to the rhythms of nature. The unified, integrated, community-based network makes more likely the adoption of sustainability as a core public policy value given its central concern with the idea that ecosystem health is essential to long-term economic and community health. The collaborative, deliberative, participative institutional dynamics, along with outreach efforts, are making more people aware of the value of ecological services, biodiversity, and more environmentally sensitive practices. Others are being convinced to use fewer resources and engage in more efficient land-use practices through projects that provide private benefits for individual landowners. Finally, the explicit reliance on a broad knowledge base facilitates innovation, problem solving, and effectiveness—all within the context of the sustainability-oriented "environment, economy, and community" mission.

8

Accountability and Policy Performance through Governance *and* Government

Critics . . . say we locals will be out-sophisticated, out-maneuvered, and thus controlled by industry, government, or national environmental groups. We think that to be arrogant, condescending, and colonial. We are not country bumpkins. We are not hard-edged environmentalists. . . . Nor are we industrialist lackeys, either. We are knowledgeable and savvy community-minded people with strong stewardship values who have long-term experience, investments, interests, and commitments to the health of our ecosystems, communities, and economies, all three of which are intricately tied to the health of the other.
—Dan'l Markham
Executive Director, Willapa Alliance
1997 U.S. Senate testimony, p. 49

Better-informed, hyperactive publics who must live with the day-to-day consequences of public policy decisions are now demanding decision-making arrangements giving them more creative control over policy, especially over how that policy will be achieved (Inglehart 1997, 232; Kemmis 2001; Sirianni and Friedland 2001). Accommodating such demands is likely to require greater use of new governance arrangements that empower citizens and stakeholders at the state and local levels of government. Grassroots ecosystem management is emblematic of this ongoing transition to alternative institutions in environmental policy, particularly in terms of the focus on collaboration, citizen participation, and the longer-term goal of a sustainable community. Yet GREM, and other similar efforts, are unlikely to make much headway as legitimate alternatives to existing institutions unless concerns over democratic accountability and policy performance are properly addressed.

The exploratory study of the three exemplary cases of GREM—the Applegate Partnership, the Henry's Fork Watershed Council, and the

Willapa Alliance—is a first step in this direction. The empirical evidence from these cases suggests that it is possible for these new governance arrangements to help resolve environmental conflicts, solve environmental problems, and ensure broad-based, simultaneous accountability when power has been decentralized and shared with the private sector, when the decision processes are premised on collaboration and consensus, when citizens actively comanage issues affecting public lands, and when broadly supported results are key to administrative success. The results also suggest that the collaborative, participative decision-making dynamic contributes positively to participants' quest for sustainable communities. Finally, the empirical record suggests that participants are redefining and broadening the idea of policy or outcome success to include an explicit focus on the accountability "to whom" question in addition to the more traditional focus on accountability "for what." In other words, the two questions of "for what" and "to whom" are seamlessly woven together. The idea is that "for what" cannot become a reality until and unless the answer to the "to whom" question satisfies a broad-based, diverse coalition of citizen-participants. In short, the answer to the "to whom" question must reflect a more holistic conception of the public interest.

The concluding chapter starts with an assessment of the GREM accountability framework in light of conventional expectations and concerns regarding the ability of GREM to produce accountability. A second section recognizes that there is likely to be considerable variance among decentralized, collaborative, and participative efforts and that not all will operate in the same broadly accountable fashion. Given this, I offer some observations about lessons learned, particularly about the factors that seem most significantly associated with the successful achievement of broad-based accountability for decentralized, collaborative, and participative governance arrangements. In addition, there is consideration of how GREM fits within the existing framework of government as a potential hybrid or supplementary model of governance, one that combines governance *with* government.

The concluding discussion is mindful of the limits associated with exploratory case-study research. It recognizes that the research in this book is not definitive, but rather is a necessary first step for future research designed to build the kind of explanatory theory that will help us to know with greater certainty whether these new governance arrangements consistently produce positive-sum environment *and* economy outcomes. In other words, is broad-based, simultaneous accountability dominant across cases, or is the typical outcome more likely to be the Bambi-versus-Godzilla scenario proffered by critics? To take us further along the road toward explanatory theory, some promising avenues of additional research are suggested that may help to confirm the findings in the case studies, or suggest how they should be modified.

Conventional Expectations and the GREM Model of Accountability

Demonstrating that broad-based, simultaneous accountability is possible is not the same thing as saying that participants have designed a perfect system of accountability. Like prior systems of accountability, the model of accountability deriving from decentralized, collaborative, and participative governance arrangements raises a variety of concerns.

Networks as Weaker Vehicles for Social Action

Milward (1996, 79) argues that "networks are weaker vehicles for social action" given the coordination problems stemming from the fact that all activity is jointly produced. The concern is one of performance; collaborative administration necessarily produces fewer decisions than specialized, strong hierarchies, and is unable to keep up with the demands for action by political and administrative superiors. The accountability problem here is fundamentally one of quantity—requisite duties are performed and compliance is forthcoming, but at a slower rate. This perspective thus assumes that quantity, or action for the sake of action, is the primary measure of accountability.

Yet what if the differing character of outcomes is taken into account? Strong hierarchies are capable of strong action (because of less need for coordination) within their relatively narrow, specialized spheres. The problem is that such specialization often translates into an incapacity to see the big picture, thus leading to outcomes that shift, rather than solve, problems into someone else's sphere of influence, or that are suboptimal or even redundant from a collective perspective. The community- or place-based network being fostered by GREM, on the other hand, coordinates and integrates first and then applies consensus solutions. Thus there *is* likely to be less action (fewer decisions), but more appropriate action that not only is supported broadly by citizens (agency representatives included), but that also provides a better fit for the realities of the collective, watershed scale of action. Given the shift in focus from quantity to quality, or kind of outcome, it is not clear that networks are weaker vehicles for social action.

Leaving aside the issue of outcome "character," however, the fragility of collaborative, participative arrangements is hard to avoid. The open-access, norms-based, consensus-oriented institutional framework poses a difficult problem for the new governance arrangements. Such efforts need to be open to all to fight accusations of special interest governance, but a dilemma arises when any contingent of participants, for whatever reason, strategically uses the rules and processes governing the proceedings to erect roadblocks and, more generally, to abuse the spirit of the effort. The refusal to "reason together" with others in search of solutions or to abide by governing norms may be enough to hamper effectiveness to the point of institutional failure. Moreover, there is little political risk or cost in such strategic behavior for those who occupy the extreme ends of the political spectrum, whether right or left. This is because success in stopping nonpreferred compromise initiatives earns them accolades among their supporters as defenders of the faith, while failure to stop others from moving ahead without them allows them to cry foul, to claim that GREM is nothing more than special interest government.

Linking Rights with Responsibilities

An institutional framework that links rights with responsibilities also troubles some critics. According to this argument, placing conditions on participation, hence on the opportunity to influence outcomes, and reserving the right to sanction nonconformity with exclusion or by discounting preferences, violates the rights of dissenting individuals. Yet unless we conceive of political rights as the ability of individuals to behave in whatever manner they deem appropriate and to obstruct decision-making processes until and unless they get exactly what they want, it is hard to see how rights are being violated. The opportunity for participation, hence influence, is still available.

In addition, people are excluded from participating in political venues all the time for violating the rules and norms governing participation. In some states, voters are denied their "right" to vote if registration procedures are violated, including a failure to fill out voter registration materials in a timely manner. Citizen-initiated ballot proposals are denied a place on the ballot if supporters fail to gather enough valid signatures. On occasion, citizens are forcibly removed from city council or university regents' meetings when their behavior prevents discussion and completion of agenda items. And preferences are discounted, too. Elected officials tend to spend much more time with representatives from organized groups than with individual citizens, or more time with members of some groups than others. Regulators have been known to place more weight on the public comments proffered by major stakeholders than by unaffiliated, individual citizens (Weber 1998). In fact, a number of participants are convinced that agency bureaucrats regularly discount their preferences for policy as expressed in formal public hearings. Finally, nothing in the new framework stops dissenters from seeking redress of their grievances in other forums such as the courts, or through duly elected officials in the different branches and levels of government.

Too Much Discretion?

Another concern arises in the area of bureaucratic discretion. The collaborative, intergovernmental character of GREM and the give-and-take of the deliberative dynamic suggest that success requires greater discretion for the agency line personnel who directly interface with the communities in question. Success also seems to require that affected line personnel have both the capacity and the opportunity to build the kinds of interpersonal relationships and individualized trust with the larger community that appear to be a key part of the glue holding the new governance arrangements together. Added discretion and the apparent need for long-term appointments raise the fear that agency employees will eventually compromise their loyalty to their parent agency, or even switch it to favor community over agency interests. The fear that agency personnel will "go native" and become captured by the community is, of course, an age-old one (see Kaufman 1967).

Governance by the Self-Selecting Few

A further concern revolves around the number of citizens participating in GREM. Because GREM is grounded in participation and direct access by community residents, there is the added accountability burden associated with attracting and maintaining the involvement of a sizable number of agency and advocacy group representatives, community residents, and so on. In each of the three cases, although the core groups of participants are relatively balanced among the major stakeholding groups of each community, there are fewer than 50 to 100 people, out of a potential of thousands, regularly engaged in decision making (although many more lend their support by volunteering on an ad hoc basis to assist implementation). Is there a threshold or magic number to satisfy this concern? How many is enough?

A related concern involves the decline over time in participation rates for national- and state-level bureaucrats evident in the Applegate Partnership case. Why did agency officials participate less over time? And what does this imply about national accountability, particularly in the absence of representatives of national interest groups? Do governance arrangements exhibiting this pattern tend to produce outcomes that are less broadly accountable vis-à-vis efforts with strong participation by federal and/or state officials or national environmental advocates?

Beyond the issue of the sheer numbers of participants, there are related objections to the idea of governance arrangements that enhance bottom-up accountability, but primarily for self-selecting, "active" publics— those willing to take the time and effort to have their voices heard. Chances are the regular participants will be elites, better educated with higher incomes and greater argumentation and persuasion skills. Others will struggle to overcome the barriers to, and the fear of, participation, whether it is time, a lack of resources, or something else. This criticism assumes that a successful accountability system requires adherence to a rigid, formal standard, namely: if direct participation is one of the keys to accountability, the only way to articulate individuals' preferences is to have each person express their views in a direct, unfiltered manner. It also assumes that adding more voices to the deliberative mix will always introduce different policy preferences. Thus, small numbers of participants equal weak accountability precisely because only a select few are making decisions.

Some of these barriers to participation, of course, can be alleviated through measures like paid participation or more flexible work schedules to allow meeting attendance, although neither is likely to enjoy wide support given cost and productivity considerations, among other things. At the same time, however, the ideal world of the critics gives little or no credence to the idea of intermediary representation by group leaders, whereby enough trust exists that at least some citizens voluntarily give their participation and decision-making proxies to others.[1] Nor is there likely to be room in such a view for participant norms that regularly stop the discussion to ensure that the views of absentee (nonattending) citizens are duly considered, albeit imperfectly, before conclusions are reached. Yet both are mechanisms that, if applied, minimize the accountability problem associated with limited numbers of participants. There is also the idea of critical mass in an iterative decision-making setting. It is generally acknowledged that a group of decision makers can bring more knowledge to bear on problem-solving exercises than a single person can. Moreover, a group of people interacting over time through multiple discussions will enjoy greater opportunity to discuss and explore possible solutions than a small handful of people in a one-time setting. Is there a critical-mass threshold of people in a deliberative, ongoing decision process that reasonably approximates a much larger array of policy preferences, perhaps even an entire community of tens of thousands of people (or more)?

Unclear Lines of Authority—Who's In Charge?

Other critics point out that a decentralized, collaborative, and participative venue, by definition, makes assigning blame more difficult, if not impossible, when things go wrong. After all, this line of reasoning goes, if all are in charge, perhaps no one is in charge, hence no one is accountable.[2] This concern derives from the clear-lines-of-authority dimension of the Progressives' model of bureaucracy. The problem is that GREM possesses no clear lines of authority. There are several ways to address this concern, although admittedly none are likely to satisfy ardent defenders of the administrative status quo. First, the lines of authority in traditional hierarchical, specialized bureaucracies are not nearly as clear as many claim. The process of thickening over time (Light 1995), and uncertainty over which political institution (West and Cooper 1989–90) or congressional (sub)committee (Hammond and Knott 1996) is the appropriate political principal, exemplify how muddied the lines of authority for administrators can be.

Second, it is not clear that the traditional blame game, where an advocacy group or a particular elected official finds a deficiency and exposes it, is oriented toward broad-based accountability in view of the compartmentalized design of bureaucracy. Rather it is plausible to argue that such actions are designed to promote or protect a particular, narrowly construed interest. The point is that clear lines of authority allow for blame, hence accountability for something gone wrong, to be placed squarely on someone or some "official" entity, but without exploring whether the accountability dynamic is narrow, broad, or something else entirely. The logic behind the argument accepts that the source bringing the problem to light is automatically concerned with broad-based accountability when that is not necessarily the case (i.e., the question of "to whom" is assumed away by virtue of the authority vested in the source of the complaint). Defenders of the traditional public administration paradigm often invoke the same argument when criticizing new public management scholars. When it comes to democratic accountability, it is not enough for government to be responsible only to certain collections of interested stakeholders, it must be responsible to the entire polity. Otherwise, the system is "by definition" unaccountable.[3]

Third, there are those who think clear lines of authority should not be violated for any reason or that the blame game–fragmented bureaucracy dynamic is not as problematic as the previous point allows. Adherents of this view can take some comfort in the fact that with many GREM programs, implementation is done by an existing government agency. In these cases the hierarchy and traditional "blame" dynamic remain intact.

Fourth, there are also historical examples, such as the case of Robert Moses in New York, or J. Edgar Hoover, director of the Federal Bureau of Investigation, where clear lines of authority and efficiency are obvious. But for what purpose? The cases of Hoover and Moses are perhaps best remembered as cases of imperial, largely uncontrollable bureaucrats either indifferent to the broader public interest, or equating it—the broad public interest—with their own narrow vision of "good" public policy (Caro 1975; Knott and Miller 1987).

Fifth, recall that the alternative governance arrangements examined here are open and diverse, and that there are a significant number of political watchdogs, particularly national environmental groups and their allies, waiting and watching to publicize every misstep and failure. The combination enhances the probability that mistakes will be noticed and blame assigned to the collective entity. So while a mistake or two, unless of grave consequence, is unlikely to bring retribution, a consistent pattern of "failure" is bound to hurt legitimacy and increase the likelihood that participants will withdraw support as well as making funding harder to come by, thus crippling the effort. As one environmentalist-participant notes: "Success breeds success. If you do a crappy job, you're guaranteed one thing—that you'll go out of business" (personal interview, 1999). In short, while the blame is not formalized, the direct, potentially cata-strophic nature of the accountability mechanism, the importance of a reputation for broad-based accountability to long-term effectiveness, and the advisory status create substantial incentives to either correct the problem expeditiously or mount a credible defense against the original charges.

Voting and Accountability

The Willapa, Henry's Fork, and Applegate governance arrangements lack an electoral system for letting the voters decide whether to continue sup-port or "throw the bums out." Voting has long been entrenched in the lexicon of American politics as essential to accountability, and it does do a better job of involving a larger number of citizens than institutions grounded in direct participation. Voting as a form of accountability, how-ever, is not without flaws. To the extent that voter turnout is high, say in the 70 to 90 percent range of all eligible voters common to many de-mocracies, it is easier to claim that voting is an effective accountability mechanism. Contemporary American politics does not enjoy such robust levels of turnout and, in fact, has experienced a serious decline in voter participation over the course of the twentieth century. Presidential elec-tion years typically draw a voter turnout in the range of 50 to 55 percent of all eligible voters, while off-year or midterm elections tend to attract far fewer voters, generally in the 38 to 45 percent range. In the case of midterm elections, assuming a 40 percent turnout, winning candidates with a bare majority of the vote are assured of two things. One is multi-year tenure in elected office and the other is the knowledge that roughly four out of five people (80 percent of the electorate) either did not vote for the winner or voted against the winner. The weakness of voting as an accountability mechanism is even more pronounced in local elections or special elections, where turnout regularly hovers around 10 to 30 percent of the eligible electorate.

Another concern with voting is whether voters are informed. In other words, what does the average voter know? What is their level of understanding about candidates? Substantive policy issues? General political knowledge? To the extent that the general electorate is relatively ignorant on issues, candidates, how the political system works, or, for example, which party is more conservative (and by default which is more liberal), the accountability relationship is weakened. Research conducted by Delli Carpini and Keeter (1996), among others, concludes that there is considerable ignorance in each of these areas.

Further, how useful is voting as a communication mechanism between citizens and elected officials when it comes to deciding policy issues? What, exactly, are voters saying in terms of policy when they elect someone to political office? American campaigning generally does not involve candidates explicitly rank ordering their policy priorities in terms of importance, much less in the sequence in which issues will be approached/solved. As a result, voting is a poor mechanism for translating the aggregate desires of the electorate into substantive choices about either policy or the means of implementation.

The Fear of Bureaucrats as Policymakers

As a corollary concern, critics are bound to point out that the new governance arrangements violate a cherished principle of American democracy by stripping elected officials of at least some of their authority to decide policy or to determine what results are going to be produced. In their place as policymakers are unelected bureaucrats, a group of neutral officials who are supposed to faithfully execute policies passed down by elected officials, not make policy themselves.[4] Behn (1999, 143) best articulates a response to this concern:

Although it has been true under the rules of the traditional public administration paradigm that civil servants are not supposed to make policy, they often do. This is, of course, the dirty, little secret of public administration. Civil servants do make policy. Typically, they disclaim that they are doing any such thing. They insist that they are merely filling in the administrative details of overall policies established by the political process. For over a hundred years, we have continued to maintain the fiction that civil servants do not make policy. It is a most convenient (though precarious) fiction. For once we confess to the unpleasant reality that, for civil servants to do their job, they *must* make policy decisions, we have to discard the public administration paradigm. Yet, by continuing to publicly profess both the principle and the practicality of the politics-administration dichotomy, the advocates of traditional public administration are able to offer an internally consistent (if disingenuous) theory for the implementation of public policy.

Despite these potential drawbacks, it is important to remember that skepticism about new governance arrangements is a time-honored tradition in American politics. For example, during the early years of the Progressives' attempts to implement civil service reform dependent on unelected bureaucratic experts, opponents railed against the idea as one that was both undemocratic and utopian (Skowronek 1982, 47). Yet the civil service idea is now entrenched to the point that it is considered an essential part of American democracy. More recently, Robert Reich (1990b), in an edited volume, has criticized dominant models for discovering and implementing the public interest (e.g., bureaucratic expertise; pluralist process model) as subordinating the kind of democratic deliberation needed for developing public policy responsive to a broad public interest. And the reality is that acceptable systems of democratic accountability have taken a variety of forms through the years rather than adhering to some sacrosanct overarching notion of accountability, especially with respect to how to achieve it. In earlier work (Weber 1999c), I identified and described five different models of accountability—Jacksonian, Progressives/New Deal, public interest–egalitarian, neoconservative efficiency, and GREM. Each emphasizes distinctive arrays of institutions and locates authority for accountability in differing combinations and types of sectors (public, private, intermediary), processes, decision rules, knowledge, and values.

Equally important is the realization that there is no perfect model of accountability. There are differences, sometimes dramatic, between the promise of accountability in the ideal and the reality of accountability in practice for the major accountability systems employed in American politics over the past 200 years (Burke 1986, 2; Weber 1999c, 483). The Jacksonian model evolved into corrupt, narrowly responsive political machines that used selective favors—favors that were excludable and divisible, rather than public in character—to exclude large segments of the citizenry from the receipt of government goods and services (Knott and Miller 1987). The principles of administration favored by the Progressives and

New Dealers promised neutrality, efficiency, and accountability. Yet rather than being a neutral conduit for policy, this model inevitably favored some interests over others in the same way its Jacksonian predecessor did. The large degree of discretion afforded to agency experts through the purposive delegation of authority, deference of the courts, and the ad hoc, "inevitably incoherent" character of legislative oversight[5] often resulted in industry's capture and control of public decisions for narrow, private purposes (Bernstein 1955; McConnell 1966), in "imperial" bureaucrats (Caro 1975), and in professional capture of public agencies.[6] Application of the public interest–egalitarian model that emerged in the 1960s and 1970s mimicked the pattern of prior historical eras with respect to the guarantee of democratic accountability. The institutional structure—federal control, hierarchical bureaucracy, an activist judiciary, a micromanaging Congress—and procedures used to ensure accountability not only meant that some groups and policy interests were favored over others, but that a series of other accountability problems accompanied the top-down arrangements.[7] Nor, as Light (1995) and Behn (2001) make clear, does the adoption of hierarchy as a principle for organizing and controlling bureaucracy necessarily guarantee accountability. In short, there is credible and mounting evidence within the scholarly literature that severe accountability problems do accompany existing governance regimes and that they exist in much more than anecdotal fashion (i.e., they are institutional pathologies) (e.g., Allison 1971; Knott and Miller 1987; Light 1995; Mosher 1982; Nelson 1995).

Getting There: What Matters Most to Broad-Based Accountability?

Perfect or not, efforts to bring society back into the administration of environmental policy, and public policy more generally, continue to emerge in communities across America. It is a given that all these new governance arrangements will not possess the same capacity to achieve a form of democratic accountability that we can label broad-based and simultaneous. The three cases of GREM, however, offer a good vantage point for ascertaining the factors or conditions likely to enhance the probability that decentralized, collaborative, and participative policy administration arrangements will achieve broad-based, simultaneous accountability. Important factors are found in the larger institutional setting within which such alternative institutions occur, and in the institutional

Table 8.1
Factors associated with broad-based accountability

Factors of first-order importance	Factors of second-order importance
Context	*Context*
1. National superstructure of environmental law and natural resource bureaucracies	1. Hetcrogcncity of community
2. Place/community dependence on nature	2. Availability of external watchdogs
Institutional mechanics	*Institutional mechanics*
1. Transparency	1. Infusion with value of broad-based accountability
2. Infusion with value of broad-based accountability	• Committed leadership
• Participant norms	• Formal decision-making procedures
• Holistic, integrated approach to public policy	
• Frequent (regular), iterative, deliberative decision process	
• Virtuous citizen-participants	
• Consensus-decision rule	
3. Credible commitment to community orientation by public agencies	
4. Diversity of representation	
5. Verifying results (regular accountability reports)	

mechanics specific to the effort (see table 8.1). Given the organic, informal character of these new arrangements, however, the true effectiveness of many of these factors will only become evident over time as the practices of cooperation evolve and relationships among participants develop.

The Larger Institutional Context

The existing superstructure of national environmental laws and natural resource agencies creates a "Catch-22" situation and provides an overarching protective framework for the environment. Federal regulations are a Catch-22 because, on the one hand, the problems and the conflict associated with top-down, fragmented federal control are a key catalyst behind the emergence of GREM, while on the other hand, if the existing superstructure of law and bureaucracy disappears, industry might not participate. This is because the protective superstructure creates the equivalent of a "2,000-pound gorilla" for environmental advocates engaged in decentralized, collaborative efforts; they can use national laws as a backstop and a reminder to others that there are other more costly and restrictive alternatives available. A case that drives the point home is the April 2001 decision by the U.S. Bureau of Reclamation in the Klamath Falls, Oregon, area to cut off all irrigation water to 170,000 acres of farmland in order to save suckerfish and coho salmon, two species listed under the Endangered Species Act. In the summer of 2001, land that had been farmed for four generations using irrigation water dried up, crops withered, and local food bank supplies dwindled, while requests for mental health services rose 60 percent during the first six months of 2001 (Welch 2001; Locke 2001). Such catastrophic consequences, while not yet the norm, nonetheless serve as vivid examples of what can happen. The superstructure component thus not only helps to ensure the viability of such collaborative efforts by giving key stakeholders a reason to engage, but also helps to ensure broad-based accountability by maintaining an institutional "safe haven" (a minimum floor of environmental protection) and increasing the likelihood that a broad cross-section of interests are at the table.

The degree to which broad-based accountability occurs may be connected to the heterogeneity, or the presence of a broad cross-section of interests, *within* the community or place in question. To the extent that there is a balance of power among environmentalists, recreationists, commodity interests, and the like, the expectation is that outcomes will more likely comport with a broad public interest. Some might carry this reasoning further to say that before power is devolved, a place's heterogeneity needs to replicate relatively closely the balance of interests in the national polity and/or that the definition of interests needs to encompass socioeconomic stratification and racial classification, especially those without organized representation. Whether these are necessary conditions are debatable and will require further testing. They also may be made moot by the diversity-of-representation factor (part of the "Institutional Mechanics" section below).

For instance, in cases where a place has a relatively homogenous population, the range of participants in GREM can still be diverse and balanced. Or the opposite can be true; there is a heterogeneous population, but participation in the new governance arrangement is tilted heavily to favor some interests over others. At the same time, there are likely to be exceptions to the rule. For example, as far-fetched as it may sound to some, certain places may be dominated by commodity producers with a progressive bent toward environmental sustainability. A blunt measure, such as proportionality of "interests," is simply unable to capture the inevitable variation in how individual members of interest groups approach the question of environmental protection, commodity-extraction practices, and so on.[8] Nonetheless, to the extent that heterogeneity within a place is lacking, it could provide an early warning sign that broad-based accountability is less likely. Policymakers and bureaucrats should then give greater weight to other factors, such as transparency, the availability of external watchdogs, and the diversity of representation in the actual effort prior to engaging the GREM format.

External "watchdogs" in the press, the general citizenry, the national environmental community, and the broader interest group community increase the capacity to monitor decisions and report on, as well as challenge, narrow decisions that have negative impacts on broad public interests (Baumgartner and Jones 1993; Heclo 1978). In fact, it is fair to say that the capacity for sounding the alarm and exposing wrongdoing is far greater and more likely to be employed today than at any time in U.S. history. Participants in GREM realize this. They know that they are being watched and that, as one local environmentalist puts it, "it's almost impossible to fly underneath the radar of national environmental groups" (personal interviews, 1998 and 1999). To the extent that this contextual condition remains unchanged, there is a greater degree of certainty that broad-based accountability will be forthcoming. However, the effectiveness of external monitoring is likely to be dependent on the degree of transparency (see below). High transparency makes it easier to monitor, while low transparency leads to the opposite effect. Moreover, as in the case of community heterogeneity, the importance of the external-watchdog factor diminishes if regular accountability reporting processes are utilized to serve the same role.

The Henry's Fork, Applegate, and Willapa cases suggest that "place" has to matter to the community or communities involved in order to achieve broad-based accountability. As noted in chapter 3, participants agree with Kemmis (1990, 67, 78) that place is a key catalyst for self-governance because it helps to mobilize citizens to care enough to participate in the act of governing "their" place by constantly reminding community members of what they have in common—the direct, tangible connection to and reliance on the natural landscape. The commonality, "or mutual stake in the shape of one another's lives" (66), enables the cooperative practices and public discussions leading to creative win-win compromises benefiting the whole of the community rather than its individual parts. At the same time, it appears that place is inseparable from a community's dependence on nature *at least when the objective of the governance arrangement is a sustainable community.* The limited number of cases, and the fact that one of the key characteristics in the three cases of GREM that comprise this study is an economy heavily dependent on nature (i.e., more than 25 percent of gross domestic product associated with commodity production and extraction as well as with amenity services like tourism and recreation), make it impossible to know with any certainty the degree to which "dependence on nature" matters. Yet it may be that the commonality derived from the direct, steady contact with the productive services rendered (or withheld) by nature, when combined with the specific cooperative practices (described below) of inhabiting one of these rural places, is a significant contributor to the overall capacity for broad-based accountability (Kemmis 1990, 81, 105).

Moreover, while Kemmis (1990), Snyder (1990), McGinnis (1999b), and others couch their arguments in the language of political philosophy and bioregionalism, there is another, similar line of argument, namely, that those without direct reliance on the landscape are becoming less and less able, or more unwilling, to recognize the costs to and benefits from nature deriving from their own lifestyle decisions. Supporting "zero-cut" initiatives in national forests, or, as Seattle politicians did in the fall of 2000, voting in favor of removing the four "salmon-killing" dams on the lower Snake River, is relatively easy when virtually all the costs are borne hundreds or thousands of miles away, or are hidden in incremental increases of homebuilding costs (which might not matter, especially if you have no plans to build another home).

Citizens in urban, or even suburban, areas "disconnected" from nature thus may be more likely to favor outcomes that disenfranchise those with natural resource livelihoods and/or that ignore or discount the complex interconnectedness found in natural systems, opting instead for decisions that displace rather than solve problems. Both scenarios violate the broad-based accountability premise, the first by disenfranchisement of part of the larger community, the second by thinking only in terms of the local, urban/suburban community. In other cases, "disconnectedness" may mean new settlers in rural areas, fresh from urban surroundings, struggling to connect the dots between resource use and resource extraction (Riebsame and Robb 1997). Participants in GREM have a number of stories in this vein, among which is the story of a new resident in the Applegate Valley, fresh from one of California's many urban centers, who built a roughly 7,000-square-foot home next to national forest land (BLM) using large, exposed timbers and vaulted ceilings and floors covered in natural wood. The resident then used the home for meetings to battle against BLM's attempts to selectively cut "his" forest behind his new home as part of a larger fire-hazard reduction plan. A federal official and participant in the Applegate Partnership lamented that "despite the fact that [the resident's] own house had taken a small forest to build and, given the size of the timbers, probably used wood from old-growth forests, too, he was completely unwilling to allow the forests surrounding his own house to be cut for any reason whatsoever" (personal interview, 1999).

Institutional Mechanics
Transparency is an absolute requirement for any decision-making arrangement aspiring to broad-based accountability. Open access to proceedings and public information systems are signs that that a governance effort takes democratic responsibilities seriously. Further, to the extent that participants develop and make use of a set of indicators for accurately[9] measuring trends as well as progress toward public goals, more information for assessing government performance is available. Individual citizens and/or organized advocacy groups thus have a new tool for monitoring and assessing government behavior and, ultimately, for holding government accountable.

There also needs to be strong evidence that the decentralized, collaborative, participative arrangement is infused with allegiance to the value of broad-based accountability.[10] Part of this asks if the practices associated with the effort create significant institutional obstacles (checks and balances) for political interests seeking to either impose their *individual* will on others or craft outcomes favoring a narrow set of interests or values. The list of essentials include participant norms, a holistic, integrated approach to public policy (mission and/or management approach), a frequent (regular), iterative, deliberative decision process, citizens imbued with virtue,[11] and a consensus decision rule (gives minorities veto power) (see chapter 3). The evidence from the three cases suggests that successful broad-based accountability will also likely benefit from, but will not be dependent on, other elements such as committed leadership and formal decision-making procedures (e.g., the WIRE process for HFWC) (see chapter 3).

Another factor essential to community-level accountability is the presence of a genuine commitment to a community orientation by public agencies and bureaucratic personnel. Such a commitment offers citizens a bona fide stake in proceedings and outcomes, hence real potential for influence, by accepting from the start the legitimacy of the community's role and goals, while also promising to use government power constructively rather than destructively. This will require that government participants have a clear sense of their own mission and responsibilities (i.e., the public interest as defined from the agency's perspective) before engaging a particular issue, and a willingness to go the extra mile and think innovatively ("outside the box") whenever doing so does not harm their ability to comply with agency mandates (personal interviews, 1998 and 1999). Further, agencies need to vest assigned personnel with the discretion and authority to make decisions on behalf of their organizations, and assign people who are comfortable with the idea that human relations and relationship building with community members is a critical part of their success. The clear emphasis on relationships and trust in these alternative institutions has convinced many participants that agency participants "need to stick around for a lot longer than they do now. It seems as if we finally get things figured out, we get some successes, we learn to trust each other and poof! the land manager is gone, up and transferred to another region, another job" (personal interviews, 1998 and 1999; see also Keown 1998). As a result, internal monitoring and management systems inside agencies will need revamping and strengthening,[12] while criteria related to financial rewards and career advancement will need restructuring to reflect the changing character of agency employment from frequent rotation to longer place-based tenures, given the added importance of relationships and trust to governance success.[13] Granting additional discretion and authority to the line personnel involved in community-based collaboratives also suggests the renewed importance of ethics training for field personnel.

There are likely limits to the ability of decentralized, collaborative, and participative governance proceedings that rely heavily on informal institutions, transparency, and so on, to produce broad-based accountability, especially over the long term, without the active presence of diverse interests. At some basic, almost visceral level, broad-based accountability seems to demand a reasonable array of different voices contributing to public policy decisions. And while the actual amount of diversity required to meet a broad-based accountability standard is debatable and unlikely to adhere to a single formula across governance efforts, a working definition includes at least three axes—across the place in question, across scales of governance (local, state/regional, national), and in terms of participation intensity. First, there needs to be either some semblance of balance among interests, or stakeholders, including unaffiliated citizens, in accordance with their proportionality within the surrounding communities, or some minimum threshold of participation (e.g., 5 to 10 percent) for these same interests. Chapters 4 through 6 offer a rough cut at this that captures the major stakeholding groups—environmentalists, recreational interests, commodity producers, unaffiliated citizens—for GREM. As with the previous discussion on community heterogeneity, others may want a more expansive definition of diversity that includes race, gender, and socioeconomic status. Second, there needs to be participation by more than just "locals." Representatives from the state/region and nation, especially from agencies with policy jurisdiction for a place, must be involved. Diversity will also be enhanced to the extent that broader geographic representation includes citizens and/or representatives from regional and national advocacy groups. Third, examining the consistency or regularity of the participation by agencies and interests can provide another measure of diversity that may be useful. An occasional visit, or an on-again, off-again approach by some as opposed to consistent, repeat interaction by others, may well indicate an imbalance of influence (Galanter 1974).

Moreover, as noted previously in the discussions of community heterogeneity and small numbers of potentially self-selecting participants, the analysis of diversity is unlikely to be as straightforward as counting heads and participation rates. Intermediary representation by group leaders means that one or two representatives may actually be proxies for hundreds or thousands of other citizens (e.g., Dale Swenson of the Fremont-Madison Irrigation District). The presence of participant norms designed to pay heed to the views and needs of absent "others" and the variances of preferences within groups are other caveats. At the same time, any discussion of "how many" participants should ideally come from various categories of the U.S. citizenry needs to consider the conflict between a participation standard linking rights with responsibilities that is rooted in equality of opportunity, and a standard that pays homage to equality of results (proportional equality). If opportunity for influence exists but an actor opts out, either by choice or by behavior, how much harm is done to the notion of broad-based accountability? Are there identifiable thresholds for participation by unaffiliated citizens as well as by agency and interest representatives? The conundrum may be alleviated somewhat by a system of regular formal accountability reports focused on outcomes (see below). Yet, even such a reporting system is unlikely to quiet the fears of many because, by itself, it does not herald the promise of democracy in quite the same, vivid manner as does a visible, relatively balanced array of competing interests and representatives from multiple levels of government. In this sense, while the outcomes assessment (report) may offer hard proof of (un)accountability, in the absence of balanced representation, it will likely suffer from a legitimacy deficit in the eyes of many and make it that much easier to question and disparage any claims to broad-based, simultaneous accountability.

Given the organic, informal character of the alternative institutions in question, a successful system of broad-based accountability is likely to require regular accountability reports that are cooperatively developed, employ independent, recognized experts, and use identifiable and replicable criteria for assessing the quality of accountability associated with outcomes.[14] To be effective, these reports need to focus on more than just finances and need to be more comprehensive than the current pattern of individual reporting to various external funding agencies (e.g., government agencies, philanthropic foundations). The criteria used in earlier chapters to assess accountability across different levels are a first step in this direction, albeit a rudimentary one. Such reports can be used simultaneously to assess more directly expenditure issues and the policy performance element of accountability—that is, the effects of decisions on governance goals. Regular accountability reports might also serve as a check on the behavior of agency personnel with long-term community liaison responsibilities, and to assess whether diversity in representation is as meaningful as claimed.

Developing such a reporting system may be closer to reality than many think. Behn (2001) suggests that universities, think tanks, and even stakeholder organizations themselves already have substantial capacities that can assist with the necessary reviews (see also Morley, Bryant, and Hatry 2001). The real test, however, may come in the political arena. The ultimate legitimacy of the reporting criteria and system requires, among other things, agreement among a diverse array of stakeholders as to what constitutes accurate measures of accountability and which monitoring protocols are best suited to verifying the presence (or absence) of accountability. Otherwise, the choice of criteria will be just as susceptible to political sniping and sabotage by the losing side(s) as any other *political* decision. Put differently, deciding on results-based measurement and monitoring criteria may not be the rational, technical exercise that many expect; instead it may be more about politics and social relations (e.g., trust) than anything else.

Moving toward a Hybrid Model? The Idea of Governance *and* Government

The new governance arrangements are more of a supplement or complement to existing institutions than a complete replacement for them. Participants themselves do not talk about GREM as a way to undermine or replace federal- and state-level controls; rather, they see their efforts coexisting with current arrangements. Part of this stems from the belief that strong public support for environmental protection makes it impossible to roll back the superstructure of federal environmental laws. In addition, GREM is advisory in nature; participants are forced to recognize and work with established centers of power. And despite proponents' attempts to reform and improve the effectiveness of existing government institutions, they continue to value these institutions as sanctioning authorities of last resort, as visible reminders of the high costs of doing business without GREM, and as an institutional "safe haven"—a minimum floor of environmental protection.

Seen from this perspective, it may make more sense to characterize these cases of GREM as a hybrid model for governance, an arrangement that combines governance *with* government in order to achieve better governance performance and enhanced accountability to a broader array of interests. There are two main areas where GREM strengthens or complements the existing system of government—the capacity for problem solving, and democracy more generally.

Problem-Solving Capacity

Unlike the responses to the deficiencies of earlier governing arrangements, most notably the political machines of the nineteenth-century Jacksonian era, the move toward alternative institutions in contemporary America is motivated less by the problem of corruption than by the problem of performance, particularly bureaucratic ineffectiveness. Behn (1999, 113) correctly notes: "American government may not be very crooked; but neither is it very effective." GREM complements and strengthens the government's capacity for problem solving, thereby heightening effectiveness in several ways.

The new governance arrangements catalyze previously untapped private- and voluntary-sector resources in the service of previously established public goals. In some cases, this is new money, either from individuals or private-sector and philanthropic organizations. A case in point is the willingness of Roger Ferguson, one of the major landowners and beneficiaries of the HFWC's Sheridan Creek stream restoration project,[15] to "be a hero by being the single largest cash donor [$40,000] . . . to make sure [the Sheridan Creek project] happened, even though he was under no obligation to do so" (personal interview, 1998). In other cases, citizens volunteer labor and intellectual capital for problem-solving and implementation activities. Moreover, problem-solving and management activities of multiple agencies are integrated, thus increasing opportunities for coordination and minimizing duplication (redundancy). In all cases, the net result is government more likely to maximize the bang for the buck, either by achieving the same goals with fewer resources, or by achieving more public goals with the same level of public resources.

In addition, the adoption of the holistic "environment, economy, and community" mission statement and a preference for a holistic, integrated approach to management are designed to avoid the spillover and reverberation effects common to the specialized and single-issue approaches of existing government bureaucracies. The problem of narrowness, or failure to see the big picture, associated with the traditional fragmented approach to management is reminiscent of a statement attributed to Wernher von Braun, the famous rocket scientist, who said "I only care if the rocket goes up. Where it comes down is not my problem." Participants in GREM, however, care where the rocket comes down; the management approach explicitly recognizes and takes responsibility for the trade-offs inherent in policy administration.

The collaborative, participative design of GREM facilitates and encourages the sharing of information and different perspectives among the many participants. Information sharing, in turn, creates opportunities for the development of more detailed and comprehensive information about the ecosystem and about economic and social conditions. The new information allows for consideration of a more robust set of policy choices and implementation mechanisms, while also increasing the likelihood that government agencies will discover innovative solutions that otherwise are beyond their reach (Dryzek 1987; Torgerson 1998, 119–121; Weber 1998, 4). The additional place-specific information can help agency decision makers make better decisions that are more likely to "fit" the actual on-the-ground conditions of the watershed in question. In short, GREM can be a mechanism for translating top-down, one-size-fits-all laws into a place-specific form without violating them.

Democracy and Citizenship

The structure and processes employed by GREM increase the likelihood that the two conditions that give *"representative* democracy . . . [a] solid footing" are met, thus helping representative democracy "avoid bankruptcy" (Kuklinski et al. 2000, 791; emphasis added). The two conditions involve the presentation and use of facts by ordinary citizens for political purposes: "Citizens must have ready access to factual information that facilitates the evaluation of public policy . . . [and] citizens must then use these facts to inform their preferences. . . . Fulfilling the first condition is a prerequisite to meeting the second; citizens can use facts only if the political system disseminates them" (Kuklinski et al. 2000, 791). The new collaborative and participative governance arrangements promote these conditions by treating information as a public right rather than as a private or privileged commodity.[16]

However, the American political system, when taken as a whole, generally fares poorly on this count. The existing political system produces high levels of uninformed citizens because key sources of political information—elected officials, political parties, advocacy groups, and the media—lack the incentive to provide the relevant facts (Kuklinski et al. 2000, 791). Politicians, political party officials, and advocacy groups regularly engage in strategic behavior and tend to present selective and biased bits of information designed to sway the electorate. The media, on the other hand, have a keen interest in presenting political news that attracts and maintains large numbers of viewers. As a result, news reporting today is often characterized as "infotainment." Campaign reporting, for example, tends to favor a "horse-race" mentality focused on who is ahead and who is behind as well as a candidate's chances for victory, rather than delving too deeply into policy specifics. In addition, specific events and personal situations that often veer into the realm of the tragic, or even the bizarre (i.e., statistical outliers far from the norm) are the currency of choice instead of reports grounded in general facts and placed in context (Iyengar 1991; Kuklinski et al. 2000). Citizens tuned into the media are also treated to a steady dose of "bad" news highlighting the problems of government, especially bureaucratic failures, with little or no attention to instances where government succeeds or "does the right thing" (e.g., Grunwald 2000a, 2000b, 2000c, 2000d—a multipart *Washington Post* series on the U.S. Army Corps of Engineers).

Of even graver concern to Kuklinski et al. (2000) is the fact that many citizens are *mis*informed rather than *un*informed. Many people not only "hold inaccurate factual beliefs . . . the wrong information—and use it to form preferences" (792), but with the spread of advanced information technologies (e.g., the Internet), the problem is growing. The problem of misinformed citizens erects an additional barrier to achieving a viable, robust form of representative democracy: "Not only does this misinformation function as a barrier to factually educating citizens, it can lead to collective preferences that differ significantly from those that would exist if people were adequately informed" (792).

Democracy also benefits because decentralized, collaborative, and participative governance institutions uncouple bureaucratic expertise, with its heavy reliance on science and scientific management, from its privileged position as the primary, sometimes the only, way to discover and decide the public interest.[17] Collaborative processes of civic discovery offset the bias of bureaucratic expertise toward narrow, technocratic solutions imposed on the polity with little consideration of the equally valid claims of those who must live with the outcomes (Reich 1990a). Moreover, by introducing local, practical knowledge into decision processes, there is less likelihood of the problematic or even tragic consequences of the formal schemes of order favored by bureaucratic experts.[18]

The lack of respect evinced in existing government institutions for local knowledge and for the preferences of nonexperts apparently is clear to many citizens, who are unwilling to accord scientific data its place as "the" unquestioned source for policy conclusions (Steel and Weber 2001). This problem is also becoming clearer to scientists: "Most scientists go into things fairly naively, thinking that as you develop information, the information will be incredibly compelling and people will say, 'Oh yeah, that's what we need to do' " (Downs 2000, 3). After interacting with local stakeholders, however, they are "like baffled parents . . . [who] often find themselves asking, 'Why won't they do what's best for them?' when farmers, fishers, or ranchers rebuff the recommendations of carefully constructed studies. Sometimes the answer is simple economics, but often it's more complicated than that" (Downs 2000, 1). Often it is about values and self-images tied to years of practice (e.g., agriculture) and intimate relationships with the local landscape (Daggett 1995; Downs 2000; Weber 2000a).

The position promoted by Reich (1990a, 1990b), Scott (1998), Mazmanian and Kraft (1999a), and others does not entail arguing for the outright rejection of science and technical expertise as determinants of the public interest. After all, recall that science, both hard and soft (social), is central to the decision processes found in the HFWC, Applegate Partnership, and Willapa Alliance. However, it is a call for supporters of a purely scientific approach to administration to recognize that practical knowledge, especially as derived from iterative, deliberative institutions, is a legitimate claimant in the struggle to define and implement the public interest. The legitimacy of the new governance arrangements stems not only from the fact that they add value by helping to decipher the complexities of social order, they also provide a needed counterbalance to the current nondemocratic bias of the bureaucratic-expertise model (Dryzek 1987; Reich 1990a).

Finally, the new governance arrangements help to reconstruct citizenship such that democracy is strengthened in a plurality of ways. GREM fosters civic virtues in individuals, offers added opportunities for active involvement in the policy process, and engages citizens in a community-building process. The combination energizes citizens, focuses them, and makes them whole through an increased knowledge of public affairs and an expanded capacity to look beyond self-interest to the larger public interest (deLeon and Denhardt 2000, 93). Moreover, as noted in the previous chapter, the institutional dynamic improves citizens' skills and changes expectations in matters related both to governing and to sustainable practices. By creating new opportunities for citizens to take control over their lives, citizens are empowered, the citizenship skills critical for self-government are strengthened, attitudes are changed, and the degree of local oversight and implementation expertise is enhanced (Landy 1993; Sandel 1996). As Kemmis (1990, 113) notes, whenever "a mediated, participatory approach to problems . . . happen[s], people find themselves being responsible for the ultimate decision, for each other, and even for their own ideologies in ways that they may never have experienced before. This taking of responsibility is the precise opposite of the move toward the 'unencumbered self.' It is, quite simply, the development of citizenship."

The problem of reconstructing democratic citizenship, or "strong" citizens, is, of course, not new—it has long been the "province of seemingly impractical democratic theorists" and therefore of limited concern to the vast majority of political science scholars, administrative practitioners, and policymakers (Barber 1984; Williams and Matheny 1995, 197). In recent years, however, given rapid economic and technological change (e.g., globalization), the political disengagement of many citizens, the power of organized interests, and growing concerns that government, in particular national government, is largely unwilling to confront and is perhaps incapable of solving the problems that confront the nation, the problem of reconstructing democratic citizenship has taken on new significance. In short, because the problem of reconstructing citizenship "is now a real problem confronting policymakers" (Williams and Matheny 1995, 197), more scholars, administrators, and citizens have become willing to entertain the possibility that strong citizens are a necessary component for successfully solving contemporary problems of governance and policy, and thereby maintaining social and political stability (Box 1998; King and Stivers 1998; Wolin 1960).

Where Do We Go from Here?

With respect to the larger debate over accountability for decentralized, collaborative, participative arrangements, whether the proponents or the skeptics are correct is an unsettled matter that needs to be investigated further. While the results in these cases are clear, they are not definitive. At this time it is not clear whether these lessons are generalizable to the hundreds of other cases of community-based environmental collaboratives across the United States. Given the apparent diversity among these efforts, we simply do not have enough evidence at this time to know whether broad-based, simultaneous accountability outcomes and win-win, positive-sum gains for the environment and economy are dominant across cases. At the same time, in view of the institutional complexities, it is doubtful that there will be a single universal framework that can be applied across cases; rather, there are likely to be several different frameworks, each relying on particular combinations of accountability "elements."

As a result, more field research and case studies are needed to document and test the new framework of accountability to either confirm the findings in the case studies, or suggest how they should be altered. The added information will help us determine with greater certainty if certain elements of the accountability framework are more important than others, if other critical elements are currently missing from the framework, or if there are combinations of variables that might lead to the same result, whether in terms of accountability or the lack thereof. In addition, firmly establishing the conditions under which broad-based accountability is most likely will help administrators and policymakers decide when to join, support, or challenge decentralized, collaborative, and participative governance efforts.

There is also a need for more systematic tests of the propositions outlined above, particularly for survey data on the nature and extent of the individual transformations, as well as on the connection between any changes and the actual GREM effort. The high numbers of participants claiming that they now give greater weight to the benefits of proposed actions for the collective whole (as opposed to their own individual interests) may well be an artifact of self-selection. In other words, participants may be predisposed toward community service and/or have a set of characteristics (e.g., demographic, attitudinal) that make it easier for them to think beyond self-interest vis-à-vis the rest of the population. As such, the support for transformational changes in worldviews found within this study cannot speak definitively to the larger issue of whether and to what degree such arrangements will be able to foment a community wide paradigm shift. It thus becomes important to ask a series of new, yet related questions.

For example, what kinds of transformations are GREM participants experiencing, and how, precisely, are these transformations connected to accountability? How effective are the efforts to inform and transform those within the community, but *outside* (nonparticipants) the GREM arrangements? How does experience with the cooperative practices of GREM relate to citizens' willingness to trust GREM as opposed to other levels of government? Does involvement translate into greater trust of all levels of government? Are citizens outside of GREM more willing to trust GREM as compared to other levels of government?

Further, we need to verify and expand our understanding of the linkages between GREM and environmental sustainability. A starting point is the development of comprehensive social, economic, and environmental baseline information for GREM communities compared to similarly situated non-GREM communities. These can be followed by long-term longitudinal studies of these same sets of communities. The seven-state Interior Columbia Basin Ecosystem Management Planning (ICBEMP) process conducted during the last half of the 1990s is an example of existing data that may prove useful as a baseline. More information is also needed on the different factors related to the added capacity for, and commitment to, sustainability. For example, how widespread and how deeply entrenched are the unified, integrated community-based networks in GREM communities? How do different levels of entrenchment affect progress toward sustainability? How robust are the linkages among citizens in the new unified, integrated community-based networks? In crisis situations, economic, environmental, or otherwise, how important is the new unified, integrated network compared to other established networks?

Given the need for additional discretion and responsibility for public land managers, further research also can help us ascertain whether there are certain personality characteristics or educational and professional backgrounds that make public personnel better suited to facilitating collaboration successfully and maintaining strong accountability to national (or state) agency goals.

Moreover, support for these alternative institutional arrangements hinges primarily on the expectation that more environmental problems will be solved; while at the same time the economic integrity and social stability of the communities in question are maintained or improved. As a result, greater attention needs to be paid to the character of outcomes and the criteria by which policy and program outcomes are assessed. How do we know a win-win-win "environment, economy, and community" outcome when we see it? Are there certain criteria to help us solve this puzzle that everyone can agree on? And who should determine the criteria? Participants in the collaborative efforts? A panel of external reviewers? Some combination of the two?

Another question involves the match between rural communities of the West and the new institutional framework that, at its core, involves a willingness on the part of those with established power and rights to share power with others in the community. Participants in the HFWC are absolutely convinced that this is key to their ongoing efforts: "None of this [the council] would have happened if Dale Swenson and the Fremont-Madison Irrigation District [the irrigators who control most water rights on the Henry's Fork River] had not agreed to share power with the rest of us. They hold the cards [i.e., senior water rights] and they decided to share power" (Jan Brown, public remarks, Eastern Idaho Watershed Conference, October 20, 1999). How many communities are there where the existing power elites are so ready to share power in such a fashion? And what kinds of incentives encourage them to share power?

What these last few questions alert us to is the inextricable, direct linkage between the conditions for successfully facilitating and sustaining collaboration over the long term, and the ability of collaboratives to attain accountability of any type. After all, if the collaborative, participative mode of decision making suffers from frequent breakdowns or completely falls apart after a few years, the question of broad-based accountability, while unavoidably important, nonetheless becomes moot if the governance arrangement is no longer operational. In fact, the problems associated with the unwillingness to observe established participant norms by some in the Applegate Partnership over the last several years has hurt its ability to arrive at consensus decisions. The difficulties are such that, as of spring 2002, partnership meetings were suspended until further notice. Over in the Willapa Bay area, on the other hand, the Willapa Alliance has been disbanded (as of fall 2000) after eight years of operation. Thus, while this research has been narrowly focused on the possibilities for a broad-based democratic accountability stemming from alternative institutions, success in achieving that result ultimately requires a deeper understanding of why collaboration emerges when it does and how it can be sustained over time. Fortunately, the last decade has seen a veritable explosion in research designed to answer just these kinds of questions, although most agree that much more is needed.[19]

Finally, there are legitimate concerns over the limits of a management approach predicated on a small scale. Locally grounded GREM efforts, although they include strong participation by state and federal government officials, may not be the best forum for addressing environmental issues that occur on a regional, national, or even global scale (e.g., global warming). How robust are the mechanisms linking these efforts to larger scale institutional efforts designed to deal with natural resource problems that are clearly national and global in scope? What steps, if any, do most GREM arrangements take to address such concerns? What steps should GREM efforts take to address these concerns?

Regardless of the answers to these questions, the movement toward reinventing government, alternative governance institutions, and sustainable communities is of more than academic interest. The new institutional matrix is fundamentally political in its effect (deLeon and Denhardt 2000, 90; Kemmis 2001; Mazmanian and Kraft 1999b; Waldo 1984). It is changing the "sets of rules of the game or codes of conduct that serve to define social practices, assign roles to the participants in these practices, and guide the interactions among occupants of these roles" (Young 1994, 3). Or, as Paul Wapner (1997, 67) argues in his study of governance in global civil society, these new "relational networks . . . can and do shape widespread behavior in matters of public concern and involvement."

Moreover, the push for reinventing government has gained significant support and a solid foothold at all three levels of U.S. government. GREM, and the watershed democracy movement more generally, are rapidly gaining acceptance as legitimate ways of managing public and private lands, natural resources, and other key issues facing twenty-first-century American communities. In addition, more and more communities across the United States are seeking to revamp governance formulas in their quest for environmental sustainability.

Given that efforts to bring society back in to the governance equation are breaking out in virtually all policy areas, are we now in an era, much like the Progressive era of 100 to 120 years ago, where significant elements of the American polity are redefining the constituent elements of a broadly acceptable system of democratic accountability? We obviously will not know the answer to this question for some time. But simply asking the question reminds us that changes in the American political landscape are almost always slow and incremental. Institutions and their accompanying dynamics, like accountability systems, are sticky; they do not change overnight. In the particular case of bureaucratic institutions, recall that the sweeping and dramatic bureaucratic reforms of the Progressives took more than sixty years before they were institutionalized at all three levels of American government.

Unsurprisingly, proponents and practitioners are not stopping to debate whether decentralized, collaborative, and participative governance arrangements will ever find an institutional niche in American politics or just a niche in the environmental policy arena. Instead they are doing their best to maintain momentum. And the tenacity and zealousness of the defenders of the new governance arrangements may ultimately win the day. A longtime observer of collaborative processes across the United States finds that proponents are pursuing their version of "environment, economy, and community" governance with a zealousness remarkably close to the religious fervor of the environmental movement during the early 1970s (personal interview, 1998). The fervor of belief in these new governance arrangements is clear in the remarks of Dan'l Markham, former executive director of the Willapa Alliance, to Congress several years ago:

Various community-based sustainable development organizations like the Willapa Alliance may come and go. But it is clear that the community-based sustainable development movement is here to stay and is a growing grassroots force to be reckoned with. . . . Communities are rising up and saying: Damn it, we are here and we are here to stay. We have a voice and we are taking charge of our destinies. We want to work with, not against others. To our friends and even our adversaries who are willing to dialogue, we thank you. And to adversaries who do not understand us and are not willing to understand us, I say, "Come, let us reason together." (Markham testimony, U.S. Senate, 1997, 52)

It is precisely this kind of passion surrounding the American West's widespread embrace of place-based, sustainable community collaboratives that leads Kemmis (2001) to argue that "Westerners . . . have finally come to the borders of a political maturity that will enable them to take responsibility for the place that made them westerners. And I believe that these commanding landscapes—these sovereign landscapes—will be best served by a people at last allowed to be sovereign over their homeland. It is precisely as a [small "d"] democrat and an environmentalist that I am convinced the West is now ready to be in charge of the West" (xii).

That said, reinventing government from the bottom up will never be easy. Yet, according to Donald Kettl (2000, 14), many more reformers are coming to the same conclusions as the proponents and participants of GREM: that the more policymakers and practitioners "have to grapple with problems that ignore traditional program boundaries, the more sense the bottom-up American approach seems to make. . . . [In fact,] in the long run, it may be the best way . . . [and] it may eventually turn out that American states and localities, without really knowing it, have charted a course for the rest of the world to follow" in the twenty-first century.

Appendix A

Interview Format

1. What do you like best about the *Applegate/Henry's Fork/Willapa* watershed/area?

2. What are the biggest threats to the quality of life in the *AP, HFWC, WA* area today?

3. How long have you been involved in the *AP, HFWC, WA*?

4. What is the frequency of your involvement in the *AP, HFWC, WA*?

5. What motivated you to join the *AP, HFWC, WA*?

6. Has participation in the *AP, HFWC, WA* led you to give greater weight to how proposed actions will affect the world outside of the watershed community? *[If response is yes, ask how and why? If no, ask why not?]*

7. Has participation led you to give greater weight to the benefits of proposed actions for the watershed community? *[If response is yes, ask how and why? If no, ask why not?]*

8. Has participation in *AP, HFWC, WA* proceedings changed your willingness to trust others? *[If yes, how and why? Also, what about former adversaries?]*

9. How are you held accountable? What is it about the *AP, HFWC, WA* arrangements that make them accountable? Examples?

10. What holds the arrangement accountable to the communities and people of the *AP, HFWC, WA* area? Examples?

11. What holds the arrangement accountable to the state and surrounding region? Examples?

12. What holds the arrangement accountable to the nation? Examples?

13. What, if anything, makes the *AP, HFWC, WA* more accountable than existing governing arrangements? Less accountable?

14. What are the sources of authority (or legitimacy) for *AP, HFWC, WA* arrangements? **For example, say that someone approached you on the street and challenged your participation in the *AP, HFWC, or WA* by charging that it was undemocratic and unaccountable. How would you respond?

15. A big criticism of locally grounded partnerships and ecosystem management efforts is that they exclude the interests of those outside the immediate area. How do you respond to that?

a. What are some ways that your efforts recognize and incorporate the concerns of citizens who live outside the *AP, HFWC, WA* area?

b. What are the *AP, HFWC, WA* obligations, if any, to the people outside the *AP, HFWC, WA* area?

16. From your perspective, how important is policy and program performance—defined as the ability of governance arrangements to deliver promised goods and services—to the overall accountability equation? Why?

17. What kinds of things do you do to ensure that agreements/programs are implemented and enforced? Who do you rely on for implementation? For enforcement? How does enforcement work?

18. Is the *AP, HFWC, WA* compatible/incompatible with state and federal efforts to do the people's business? Why? Examples?

19. Are there any steps you would take to strengthen the accountability mechanisms of the *AP, HFWC, WA*?

20. How successful do you think the *AP, HFWC, WA* has been to date?

21. What are the signs of success/examples of accomplishments?
**If not obvious in answer, prod on specifics: How has the arrangement helped to improve the:
—Environment?
—Economy?
—Community?

22. In what ways has the *AP, HFWC, WA* strengthened the capacity of the community to preserve and protect the environment, economy, and community of the region? Are there examples where the opposite is true (i.e., ways in which community capacity has been weakened)?

23. Is there anything that you think I missed or anything else you think I should know?

Appendix B

Henry's Fork Watershed Council

Participating Organizations and Agencies
(total of 60 organizations and agencies)

Ashton Area Development Committee

Ashton Power Plant/Pacific Corporation

District VII Health Department

Ecosystems Research Institute

Environmental Protection Agency

Fall River Rural Electric Cooperative

Foundation for Community Encouragement

Fremont County Commissioners

Fremont Economic Action Team

Fremont Heritage Trust

Fremont-Madison Irrigation District

Friends of Conant & Squirrel Creeks

Friends of Fall River

Greater Yellowstone Coalition

Henry's Fork Foundation

Henry's Fork Natural Resources Council

Henry's Lake Foundation

High Country RC&D

Idaho Department of Fish & Game

Idaho Department of Lands

Idaho Department of Parks & Recreation

Idaho Department of Water Resources

Idaho Division of Environmental Quality

Idaho Farm Bureau

Idaho National Engineering Laboratory

Idaho Rivers United

Idaho Soil Conservation Commission

Idaho State University, Department of Biological Sciences

Idaho Water Users Association

Idaho Wildlife Federation

Madison Soil Conservation District

Marysville Hydro Project
Nature Conservatory of Idaho
North Fork Protective Association
Northwest Policy Center
Northwest Power Services
Region VI Wildlife Council
Ricks College
Shoshone Bannock Tribes
Snake River Cutthroats
State & Local Elected Officials
Targhee National Forest
Targhee Timber Association
Teton County Economic Development Council
Teton County Planning & Zoning
Teton Soil Conservation District
Teton Valley Land Trust
University of Idaho
U.S. Army Corps of Engineers
U.S. Bureau of Land Management
U.S. Bureau of Reclamation
U.S. Fish & Wildlife Service
U.S. Forest Service
U.S. Geological Survey
U.S. National Park Service
U.S. Natural Resources Conservation Service
Water District I
Wilford Canal Company
Wool Growers Association
Yellowstone Soil Conservation District

Appendix C

Group _____ Project Name _____ Date _____

1. Watershed Perspective: Does the project employ or reflect a total watershed perspective?

The project demonstrates an understanding of the relationships that exist among:

  a. Physical parameters of watershed (soil formation and other geologic processes).

  b. Surface and ground water resources (headwaters and lowland resources).

  c. Biological components (aquatic life, plants, animals, and other species).

  d. Ecological communities (forests, meadows, riparian zones, migration corridors, nutrient cycles, predator-prey relationships).

  e. Human communities (towns, transportation corridors, historic/archaeological sites, economies).

  f. Climatic factors (weather patterns, air quality).

COMMENTS AND/OR CONDITIONS:

Yes _____   No _____

_____ NOT APPLICABLE

2. Credibility: Is the project based upon credible research or scientific data?

a. The project demonstrates use of scientific principles and procedures (rather than strictly a response to political agendas or impending crises).

b. The project clearly cites references or current research results to support its approach, or meets research goals and objectives set by the Council.

c. The project has undergone appropriate regulatory processes.

d. The project's goals and approach are clear and understandable by the general public.

COMMENTS AND/OR CONDITIONS:

Yes _____   No _____   _____ NOT APPLICABLE

3. Problem and Solution: Does the project clearly identify the resource problems and propose workable solutions that consider the relevant resources?

a. The project demonstrates that problems exist, using scientific evaluation.

b. The project contributes toward the maintenance, enhancement, or restoration of specified resources to proper functioning condition.

Cumulative effects or project strategies have been considered.

COMMENTS AND/OR CONDITIONS:

Yes _____   No _____   _____ NOT APPLICABLE

4. Water Supply: Does the project demonstrate an understanding of water supply?

   a. The project describes the quantity, quality, timing, and source(s) of water involved.

   b. The project considers potential impacts to water interests within and beyond the Henry's Fork Watershed.

   c. The project demonstrates an understanding of watershed dynamics and regional water policy

   (i.e., Snake Plain Aquifer, Minidoka Project, Columbia River Basin).

COMMENTS AND/OR CONDITIONS: Yes   No

_____ NOT APPLICABLE

5. Project Management: Does project management enjoy accepted or innovative practices, set realistic time frames for their implementation, and employ an effective monitoring plan?

   a. The project sets reasonably achievable objectives and measurable results.

   b. The timeline for project implementation is clear and has contingency plans.

   c. A monitoring plan is in place to effectively evaluate the project.

COMMENTS AND/OR CONDITIONS: Yes   No

_____ NOT APPLICABLE

6. Sustainability: Does the project emphasize sustainable ecosystems?

   a. The project recognizes the natural limits of the resources involved.
   b. The project helps to ensure the sustainability of the ecosystem for future generations.
   c. The project recognizes the importance of maintaining the basin's biological diversity, preventing the need for species listing under the Endangered Species Act.

   Yes _____ No

   COMMENTS AND/OR CONDITIONS: _____ NOT APPLICABLE

7. Social/Cultural: Does the project sufficiently address the watershed's social and cultural concerns?

   a. The project describes the quantity, quality, timing, and source(s) of water involved.
   b. The project considers community welfare, health and safety needs, and local lifestyles in its design.
   c. An understanding of ongoing social change and its costs and benefits to local communities is demonstrated.
   d. The project considers the development pressures being sustained by the basin.

   Yes _____ No

   COMMENTS AND/OR CONDITIONS: _____ NOT APPLICABLE

8. Economy: Does the project promote economic diversity within the watershed and help sustain a healthy economic base?
   a. The project creatively supports a sustainable basin economy.
   b. It is clear who benefits from and who is sharing the costs of the project.
   c. The project employs local labor, materials, and expertise where realistic and available.

COMMENTS AND/OR CONDITIONS:

Yes _____ No

_____ NOT APPLICABLE

9. Cooperation and Coordination: Does the project maximize cooperation among all parties and demonstrate sufficient coordination among appropriate groups or agencies?
   a. The project utilizes the expertise and talents of local citizens, agencies, and scientists and outlines how communication among these interests will be maintained.
   b. The project transcends political agendas and jurisdictional boundaries.
   c. The project maximizes efficiency among agencies and is coordinated with other activities in the watershed/subwatershed

COMMENTS AND/OR CONDITIONS:

Yes _____ No

_____ NOT APPLICABLE

10. Legality: Is the project lawful and respectful of agencies' legal responsibilities?

a. The project complies with federal, state, and local laws and regulations, including NEPA and ESA.

b. The project respects vested water rights and protects the beneficial consumptive and nonconsumptive uses of water established by law.

c. The project points out any conflicts in legal mandates and suggests any needed changes in laws or regulations.

d. The project recognizes both public and private property rights in its design.

COMMENTS AND/OR CONDITIONS:

Yes _____ No _____ NOT APPLICABLE

Notes

Chapter 1

1. The Quincy Library Group qualifies as a case of grassroots ecosystem management (GREM) despite the fact that the group eventually turned to the U.S. Congress for legislation to help them implement their plan. Even with the turn to Congress, the QLG still relies "extensively" on cooperative decision-making processes (GREM is defined and discussed later). See Wondollek and Yaffee 2000, however, for an extensive discussion of why some consider the QLG "an interesting anomaly in collaborative resource management" (229, note 1).

2. The money came from the U.S. Department of Agriculture budget. See Christensen 1996; Little 1997; Davis and King 2000.

3. For example, U.S. Representatives Don Young (AK), Helen Chenowith (ID), and Wally Herger (CA).

4. See Daggett 1995; John 1994; Johnson 1993, 1995b; Jones 1996; Kenney 1999; Klinkenborg 1995; Leach and Pelkey 2000; Lubell et al. 2002; Marston 1997b, 1997c; Pelkey et al. 1999; Snow 1997; Weber 2000b; Yaffee et al. 1996.

5. Daniel Kemmis used this term to describe the phenomenon in his keynote address at the Eastern Idaho Watershed Conference in Pocatello, Idaho (October 20, 1999). See also Kemmis 2001.

6. Some might be inclined to add DeWitt John's (1994) "civic environmentalism" label to this long list, although he describes policy-specific, single-play collaborative games and solutions, whereas GREM is fundamentally iterative and concerned with more than just environmental policy. *Grassroots ecosystem management* is used as a label for several reasons. First, it is an accurate descriptor of efforts to bring society back in and to focus management efforts in holistic fashion on the whole of the ecosystem (or place) as well on as the threefold mission of environment, economy, and community. "Civic environmentalism" successfully captures the first element but not the second, because environmentalism is most often associated with the contemporary environmental movement—a movement that shares some things with GREM but that is radically different in many other ways (see Weber 2000b). Second, for many practitioners and scholars, the word *conservation* in the "community-based conservation," "collaborative

conservation," and "community conservation" labels connotes a connection to the Progressives' conservation movement as represented in the persona of Gifford Pinchot, among others. The problem is that GREM, although it shares some of the conservationists' philosophy, is a distinctive environmental movement with far more differences than similarities in philosophy, institutions, and the approach to natural resources management (see Weber 2000b). Given this, "community-based conservation," or any label with *conservation* in the title, can be construed as a misleading descriptor. Third, the "watershed movement" is too broad as a label to capture the full dynamics of GREM and may include either top-down federal initiatives and/or a different, scientifically inspired spatial scale for management that may or may not include a grassroots or community component. Finally, "cooperative ecosystem management" connotes an image that comes closest to GREM but again fails to clarify the significance of the community-based grassroots element of the new governance arrangements.

7. The economic activity stemming from the amenities and raw materials from nature comprises a major portion of the economy (greater than 25 percent).

8. See Arrandale 1997; Daggett 1995; Haeuber 1996; Interagency Ecosystem Management Task Force, 1996; Kemmis 1990; Little 1997; Malpai Borderlands Group, 1997; Natural Resources Law Center, 1996; Nelson and Chapman 1995; Rieke and Kenney 1997; Rolle 1997a, 1997b; Snow 1997; Weber 2000a; Willapa Alliance 1996h; Wondollek and Yaffee 2000; Yaffee et al. 1996.

9. See Bayley 1994 and Sparrow 1994 for community policing, Sparrow 1994 for tax administration, Matthews 1996 for education, Bardach and Lesser 1996, 206–207, for elements of the federal JOBS (Job Opportunities in the Business Sector) program, Radin et al. 1996 and Radin and Romzek 1996 for rural development, and Walters 1997, 160–162, for public health (the "Oregon option" discussion).

10. For the theme of civic rejuvenation, and innovation more generally, see Berger and Neuhaus 1996; Broder 1994; Dionne 1998; Etzioni 1996, 1998; Glendon 1997; Sirianni and Friedland 2001; Skocpol and Fiorina 1999.

11. For examples, see Chertow and Esty 1997; John 1994; Knopman 1996; Kraft and Scheberle 1998; Mazmanian and Kraft 1999b; O'Leary et al. 1999; Rieke and Kenney 1997; Snow 1997; Yaffee et al. 1996.

12. It used to be that citizen participation in governmental processes was a largely passive affair (e.g., through voting), whereas administrative processes were the domain of bureaucratic experts and citizens enjoyed only limited access to administrative decision processes (Caro 1975; Knott and Miller 1987). Toward the mid-twentieth century, organized interest groups began to dominate administrative proceedings using an expanding array of participatory devices. National environmental advocacy groups, in fact, were at the forefront of the push for expanded public participation (Snow 1996). Despite substantial successes, such as the addition of new participatory procedures like public hearings, notice-and-comment rulemaking, and the Federal Advisory Committee Act of 1972, for example, the vast majority of the general public maintained its passive role (Gormley 1989; Weber 1999c). With respect to the many natural resource and

environmental agencies like the U.S. Forest Service, the Bureau of Land Management, and the U.S. Environmental Protection Agency, "elaborate public involvement mechanisms . . . were created in the image of the technocratic model that was well entrenched in agency traditions: 'Tell us your concerns, and we will figure out a solution.' This linear model of public involvement was better than no involvement at all but was rarely satisfying to those who participated. Further, it raised expectations that agencies were then unable or unwilling to satisfy" (Wondollek and Yaffee 2000, 13). In practice, this has meant that citizens shape policy either at the margins, after the possible alternatives are decided, not at all, or only from their own self-interested perspective (Beierle and Cayford 2001; Rosenbaum 1976).

In recent years, however, there has been a decided shift toward more intensely participatory and iterative institutional forms that integrally involve citizens in more deliberative formats, together with bureaucratic experts, to determine what the public interest actually is and/or what policy mechanisms are best suited to achieving public goals. The expectation is that the expansion of opportunities for direct, more substantive participation will lead to benefits for public policy and democracy, more generally. The expected benefits include, among other things, increased trust in government, increased citizen capacity for problem solving and self-governance, new information for making improved policy decisions, and, for some, the prioritization of environmental protection and sustainability at or near the top of the public policy agenda (Beierle and Cayford 2001; Berger and Neuhaus 1996; Box 1998; Dryzek 1997; Kemmis 1990; Paehlke 1995; Press 1994; Prugh, Costanza, and Daly 2000; Torgerson 1998). Equally important is the expectation that the new participatory style will "strengthen civic capacity and social capital. . . . [Participation in this sense] is viewed less as a competitive negotiation among interests than as a way to identify the common good and act on shared communal (versus individual) goals" (Beierle and Cayford 2001, 10; Reich 1990a, 1990b; Sirianni and Friedland 2001; Williams and Matheny 1995).

13. The push for the adoption of collaborative decision processes, while clearly related to the growing demands for participation by citizens, is also driven by the belief that public policy effectiveness can be improved if government shares at least some decision-making authority with others (Bingham 1986; Fisher and Ury 1981; Rabe 1994; Susskind and Cruickshank 1987). According to this line of reasoning, the pooling of resources, whether financial, information, personnel, or otherwise, with private- and voluntary-sector actors increases government problem-solving capacity, especially for the kinds of "wicked," complex problems that it is unlikely to solve on its own (Chertow and Esty 1997; O'Leary et al. 1999; Gray 1985; Matthews 1996). Part of the added problem-solving capacity involves the relationship between information sharing among the various partners and innovation: "Information sharing creates opportunities for the development of a more robust set of policy choices and implementation mechanisms; the added information permits participants to discover innovative solutions to environmental problems that otherwise are beyond their reach" (Weber 1998, 4). The additional information also can help government decision makers make better decisions that are more likely to "fit" the actual on-the-ground conditions of a

particular watershed, for example, without harming the intent of existing legislative mandates. At the same time, collaboration offers opportunities to reduce the high transaction costs associated with information search, monitoring, and enforcement activities that often accompany traditional top-down, command-and-control decision processes (Weber 1998).

Successful collaborative efforts thus can extend the policy effectiveness of government by developing solutions that get more bang for the buck, not only by getting more of a specific policy benefit (e.g., environmental protection) for the same amount of money as in the past, or getting the same amount of policy benefit for less, but also by avoiding unnecessary program redundancies and spillover effects that simply shift the negative effects of a particular policy problem to another policy realm. Dan Daggett, a soldier for twenty-two years in the "environmental wars," sums up the lure of collaboration in his 1995 keynote address to the Henry's Fork Watershed Council's State of the Watershed Conference: "To achieve [our] goals, [we] learned that [we] needed to help others achieve their goals. The old way is a win-lose conflict, more for me, less for you. The new way is cooperation, a win-win deal, where there is more for everyone" (Daggett 1995, 1–2).

14. Sustainability science and the idea of sustainable communities complement the calls for greater, more substantive citizen participation and collaborative decision processes. They do so by recognizing that top-down, expert-only approaches are inherently unable to deal with the complexity of the sustainability challenge and that long-term environmental policy success necessarily must involve the citizenry ultimately responsible for translating sustainability theory into on-the-ground results (Mazmanian and Kraft 1999a, 1999b; Hempel 1998; Nijkamp and Perrels 1994). The move toward ecosystem management, more generally, also "require[s] a much more active role for citizens than was true of past resource planning and management efforts" (Cortner and Moote 1999, 61). See also chapter 7 for an extensive discussion of sustainability science, as well as of the connections between GREM and environmental sustainability.

15. There is a certain irony in environmentalists' strident denunciations of the new governance arrangements, notes Snow (1996). It was "the very work of environmentalists themselves that opened the door to these new experiments with accountability and power. Having demanded public involvement for more than 25 years, environmentalists are now seeing one of the healthy outcomes of that demand: local collaboration groups, which now seek to implement what environmental laws have tried to mandate for a generation" (Snow 1996, 42).

16. See Cortner and Moote 1999; Leach and Pelkey 2000; Lubell et al. 2002; Mazmanian and Kraft 1999a; Wondollek and Yaffee 2000.

17. See also Radin and Romzek 1996, 81. Radin and Romzek (1996) argue that for the National Rural Development Partnership, predicated as it is on both intergovernmental and horizontal collaboration, "the more effective accountability relationships are likely to be those which afford some degree of discretion to the participants," particularly what they identify as political and professional accountability types (74). While the professional accountability type *is* traditional,

they call their version of political accountability *atypical* because what matters is the direct responsiveness to rural residents "without elected officials playing their traditional mediating role" (81). Bardach and Lesser (1996, 223) find that "the traditional accountability system . . . does not always measure up. It may provide the appearance of solidarity but the reality of hollowness."

18. See Bromley 1992; McKean 1992; Ostrom 1990; Ostrom and Schlager 1997; Young 1989, 1997.

19. Keohane and Ostrom 1995. Milward and Provan (1999, 3) argue that "the essence of governance is its focus on governing mechanisms—grants, contracts, agreements—that do not rest solely on the authority and sanctions of government."

20. Bardach and Lesser (1996) call this the idea of partner accountability. Participants will feel accountable to and be held accountable by the various partners within the collaborative. They stress that this is a potential mechanism for improved accountability.

21. See Radin and Romzek 1996 and Radin et al. 1996 for a close examination of accountability and other related issues associated with the intergovernmental, collaborative state rural development councils (SRDC) and the National Rural Development Partnership (NRDP). See also Kearns (1996), who argues that as collaborative partnerships and reliance on nonprofit organizations increasingly become the norm, there is a need for a broader conception of accountability that expands the range of people and institutions to whom public and nonprofit organizations must account (9) and for "keep[ing] the notion of accountability at the forefront of [bureaucratic and nonprofit] strategic planning and management systems" (xiv). Bardach and Lesser (1996) concentrate on the problems associated with traditional, financially based notions of accountability and examine how collaboratives can increase administrative capacity and effectiveness, yet pose problems for our "ability to impose accountability on these redesigned [collaborative] structures" (198). They posit that "new accountability systems, systems better suited to the policy and program purposes for which collaboratives are crafted, can be designed to substitute for the traditional system" (222).

22. See Romzek and Dubnick 1994; Peters and Savoie 1996. Milward (1996) notes the potential for accountability problems in networking arrangements but does not pursue the question in any empirical detail.

23. See Behn 1999. To be fair to Behn, his critical analysis of the reinventing-government focus on performance as *the* new standard for accountability provides a useful corrective for those intent on treating performance as the only way to think about accountability.

24. The cases also met access and cost criteria. All three cases lie within 600 miles of Pullman, Washington (the home of Washington State University). The close geographic proximity—yes, 600 miles is practically next door in the American West—reduced the costs of doing the research.

25. At the time cases were selected and research conducted on these cases of GREM, these cases were widely touted as institutional successes, something to

be emulated. However, two of the cases—the Willapa Alliance and the Applegate Partnership—have since fallen on hard times. As of fall 2000, and after eight years of meetings, the Willapa Alliance was disbanded. As of spring 2002, Applegate Partnership meetings were suspended until further notice.

26. Elected officials and their representatives were selected according to whether they represented the specific geographic area, their degree of familiarity with the effort in question, and/or whether they were in a position of authority related to natural resource policy.

27. Moseley 1999, 62. The definition is taken from Snyder 1990.

28. North 1990, quote taken from Young 1996, 247; also see March and Olsen 1989.

29. These new institutional arrangements resurrect an age-old puzzle for scholars and policymakers that goes to the heart of the study of public administration and public policy in a democracy: how to achieve improved governance performance and democratic accountability simultaneously. See Durant 1998; Kearns 1996, 18–19; Kettl 1996, 10; Thompson and Riccucci 1998; Wilson 1994, 668.

30. Personal interviews, 1998 and 1999. The fact that a formal political process (legislative or regulatory) has articulated a manifest public goal—a policy deemed in the public interest—that the institutions of government are unable to achieve represents a legitimacy problem for democratic institutions of governance. Accepting this scenario as *non*problematic implicitly accepts that government institutions have no obligation to deliver on promises made. Bardach and Lesser (1996, 201) echo this logic in their study of human services collaboratives by arguing that the "main value of an accountability system . . . is to motivate better performance than would otherwise occur."

31. The logic is much the same as that pointed out long ago by Woodrow Wilson ([1887] 1997, 20): a good degree of trust is required if elected officials and bureaucratic superiors are going to delegate discretion to subordinates in any administrative arrangement.

32. Mazmanian and Kraft (1999) point out that while there is a growing legion of advocates for sustainability as a way to rectify the flaws of existing institutions, the ideas remain too abstract and vague for most practical purposes. Indeed, little scholarly and systematic evidence exists on the extent to which the rhetoric is being put into practice, or if it ever can be (1999a, xiii, 1996b, 26). Nor is the paucity of empirical data limited to environmental policy or the issue of sustainability; it exists across the broad range of governance scholarship (Milward and Provan 1999, 24, but see Milward and Provan 1995 for work in this area).

Chapter 2

1. The watershed also contains a small corner of Clark County.

2. Land ownership varies by county. Madison is 25 percent public, 75 percent private, while Fremont is largely public (70 percent) and Teton is evenly split between public and private.

3. See Geddes 1998. In the *1991–1992 Green Index* (Hall and Kerr 1992), Western states (including California) ranked relatively high in terms of environmental quality, with Oregon ranking third (among the fifty states), Nevada ninth, Colorado tenth, Idaho eleventh, Washington thirteenth, Montana fifteenth, California nineteenth, New Mexico twentieth, Utah twenty-second, Wyoming twenty-fifth, and Arizona twenty-sixth. Amenity migrants have made the West and Northwest the fastest-growing regions of the United States (Riebsame and Robb 1997).

4. Willapa Alliance, 1998c, 2.

5. Nor is it clear that "just taking the animals off the land" is a cure-all solution; simply removing the animals "won't restore the land" (Daggett 1995, 9). The quotes are from Kris Havstad, after studying results of a seventy-five-year experiment on the Jornada Experimental Range in New Mexico.

6. Otherwise known as the "first in time, first in right" or, more colloquially, as the "use it or lose it" doctrine.

7. Hannum-Buffington (2000) examines six different sources of data, including community surveys and all Partnership meeting minutes from January 1993 to July 1999. Of the thirty-six issues noted by Applegate residents and Partnership participants, the issue of forest fires ranks second in overall importance, but first in terms of total number of pages of text devoted to discussing the issue (70–73).

8. Interviews with other federal agency officials confirmed that when the Applegate Adaptive Management Area Ecological Health Assessment was being conducted, citizens from the Applegate Partnership were clearly cognizant of, and pushing for programs to correct for, the declining health of area forests (personal interviews, 1999). See also Applegate Partnership meeting minutes, March 3, 1993.

9. These figures are not adjusted for inflation.

10. See Cleveland 1997, 23–26.

11. Personal interviews, 1998; *Chinook Observer* 1996; also see Willapa Alliance and Pacific County Economic Development Council, 1997.

Chapter 3

1. There are obvious limits. See the exchange between Susan and Jack Shipley (2000), original members of the Applegate Partnership, and Chant Thomas (2000), a member of the Headwaters environmental group and representative of The Threatened and Endangered Little Applegate Valley (TELAV) group. The position promoted by TELAV—zero cut, zero management of forests—is viewed by the Shipleys as threatening long-term ecosystem health because it ignores the fact that "the Applegate Valley is a 'fire dependent' ecosystem that was aggressively managed for several thousand years by the indigenous populations. [Yet, starting] seventy years ago our society decided to suppress fire in this ecosystem . . . [leading to] dense, overstocked [high-fuel-load] forests that are threatened

with catastrophic fire events" that will negatively affect the resilience and productivity of area ecosystems. The Shipleys are also concerned that TELAV is misinforming people about current BLM timber management practices and the results of proposed BLM timber sales, while also seeming to be more interested in promoting conflict and dividing the community using "scare stories" than in finding ways to collectively solve ecosystem management problems.

2. This outcome belies Morone's (1998) argument about the "new populism" in the United States. Participants adhere to "the republican faith in a shared, common interest" (328), while also failing to adopt the new populists' penchant for "split[ing] the people into us and them" (328).

3. The general literature on collaboration identifies geographic scale as one way to keep collaboration manageable. There are others; see Weber 1998, 105–119.

4. However, there are no formal restrictions against outside observation and participation.

5. Individuals certainly bring self-interest and individual priorities to collaborative watershed approaches to governance, but the commitment to "place" and the potential for transformation through participation appear to encourage people to see beyond their self-interest and to be motivating factors for joining such efforts (personal interviews, 1998 and 1999).

6. The Applegate Partnership and Willapa Alliance express the same support for existing laws in documents and interviews. See chapters 4 and 6, respectively.

7. The design and operation of HFWC meetings draw heavily from Scott Peck's (1978) *The Road Less Traveled*.

8. The inadequacy of information is due in large part to "the generally undefended, incomplete and largely incomprehensible [character] . . . of past and current monitoring systems" (Grand Canyon Trust, 1999, 3).

9. Except in cases where contracts are signed to provide specific goods and services.

10. The dynamic echoes North 1990. While "formal rules . . . make up a small (although very important) part of the sum of constraints that shape choices; a moment's reflection should suggest to us the pervasiveness [and importance] of informal constraints" (36) on human behavior. Not only do informal institutions "determine the calculus through which the all important incentives of formal rules will affect choice, . . . they provide for much of the enforcement [of agreements] essential to any [governance arrangement]" (Ensminger 1996, 181).

11. See Fukuyama 1995; Putnam 1993. Braithwaite (1998) actually takes this line of reasoning a step further. He argues that the enculturation of trust within institutions "controls the abuse of power. . . . It is [thus] the only technique for controlling abuse of power that not only averts a major drag on [governance] efficiency but actually increases efficiency" (351).

12. These events occurred during the summer of 1998. Earlier the same day of the two other events, activists from Williams also visited the USFS Applegate Ranger District offices. However, in the Applegate Ranger District "invasion"

they did not wear ski masks. Instead they walked throughout the main building and put "propaganda" on the desks of USFS employees (personal interview, 2000).

13. Glendon (1995, 6) argues that "ordinary civility may be the indispensable precondition to the acquisition of civic skills and [other] virtues."

14. He was hired as executive director in fall 1998.

15. The HFWC fits within the larger trend toward responsive regulation in which policymakers, regulators, resource managers, and resource users alike are recognizing that there "is not a clearly defined program or set of prescriptions concerning the best way to regulate" (Ayres and Braithwaite 1992, 5).

16. Fesler and Kettl (1991, 327) call this program accountability. Behn (2001) refers to it as *performance-based,* or *results-based accountability.*

Chapter 4

1. See King, Keohane, and Verba 1994, 51–53.

2. In the early years, meetings were held twice each week; since 1996, they have been held once a week. This means that in 1994 and 1995 there were 102 meetings each year (51 weeks times 2; no meetings during Christmas holiday week). For 1996 through the end of 1998 there were 153 meetings (51 per year), with another 28 meetings through the July 1999 date. The total number of possible meetings, therefore, equals 385.

3. A total of 3,633 "visits" were made to the 205 meetings, making the average attendance for meetings 17.72 people.

4. The Applegate Partnership keeps a record of attendance for each meeting that is part of their publicly available meeting minutes. Because records only include the names of individuals, further legwork was needed to establish organizational or interest affiliation, and the address or location of each individual participant. The final database was constructed by (1) compiling the complete list of names for all 205 meetings in alphabetical order, (2) asking seven current and former Applegate Partnership participants who have had high rates of participation over time and who represent a broad cross-section of interests to identify interest affiliations and "locations" of meeting participants, and (3) double-checking "locations" ascribed by the seven participants against area phone books.

5. Other bureaucrats agree. An EPA official argues that "it's always a better rule if you're touching base with all the constituencies—the environmentalists, the states, industry, the consumer groups—because more information about how it's going to work or not work is always valuable" (Weber 1998, 125). See Weber 1998, 125–140, for a more general discussion of how iterative, collaborative forums "provide a venue where the rank ordering of 'political' priorities [can] be clarified, the middle ground between contentious issues explored, and innovative compromises worked out and incorporated into the final regulatory program" (140).

6. This work was done with the assistance of John Mairs, professor of geography at Southern Oregon University (Shipley 1996, 3).

7. The goals of the Partnership were facilitated somewhat by the mission convergence brought on by the Clinton administration's Northwest Forest Plan, in particular the designation of the Applegate area as part of an Adaptive Management Area (AMA). However, the Partnership approach came first and was, in fact, considered by many to be a model for how the AMA concept might work (personal interviews, 1999).

8. See the Applegate Partnership paper from meeting minutes, "Stewardship Contracting" (1997) for a description of this philosophy. It is "not about local control," but rather about involving the local community in decisions affecting local forests: "We want three things from stewardship contracting. One is steady, well-paying jobs for our children and ourselves. Two, a means to stay in our homes and continue to live and work with the natural resources and the life we choose. Three, a voice in what happens to our lives and surroundings, so that we not be edged out of our homes, our children left without hope and the character of our rural towns lost forever" (1–2).

9. An Applegate Partnership participant erected the billboard.

10. Early estimates suggested that the cost of executing the ecosystem-restoration timber sale would be approximately $500 an acre, significantly above costs incurred for a standard commercial timber sale.

11. The Aerial Forestry Management Foundation, a group involved in the founding of the Applegate Partnership, was the first to make this proposal in late 1996 (see Moseley 2000, 25–29).

12. Landscape design is a process of examining the structures and flows of a landscape, including water resources, and determining a desired future condition based on the ecological limits of the landscape, preexisting policies, and the goals of participants (Moseley 2000, 27–28).

13. The Watershed Council is legally the same as the Partnership. See the discussion below for how this relationship works. The Applegate RD paid the ARWC $70,000 for the environmental assessment.

14. This particular effort benefited from a new state law that provides funds for regional problem-solving efforts. Without the state funding, it is not clear such a large information-gathering effort would have been possible (personal interview, 1999).

15. I draw this conclusion based on the fact that improvements in environmental quality were made. Further research is needed to determine if the cooperative solution brought the property into full compliance with state and federal water quality standards. In other words, if the changes did not result in full compliance with existing laws, it becomes *harder* to argue that there is accountability to these levels. I stop short of saying it is *impossible* to argue that there is zero accountability to these levels because it is not clear whether coercive-based solutions or sanctions would generate full compliance either (although a complete shutdown of the stockyard operations might do the trick).

16. Other examples of projects implemented through cooperation with private landowners and public agencies include (1) a community assessment (funded through cost sharing between the USFS, BLM, Southern Oregon University, and Rogue Institute of Ecology and Economy), and (2) a cooperative assessment of the ecological and economic health of the watershed involving the BLM, USFS, Oregon State University, Southern Oregon University, USFS Pacific Northwest Research Station, among others (Shipley 1996, 3).

17. Powell (1997) explains how *Babbitt v. Sweet Home* (115 S. Ct. 2407 (1995)) expanded the "harm" definition of the ESA to include habitat.

18. "It's a waste of money to manage the upper forests well, then have our efforts go for naught because in the [private property] lowlands too much water has been taken out for irrigation, or riparian areas [have been] cleared so that the water becomes too hot to sustain key aquatic species" (personal interviews, 1999 and 2000). See also a special issue of *Ecological Applications* (1996) for a series of articles on the importance of ecosystem management and the need to integrate land management across sectors, including a number of pieces by leaders and former leaders of federal agencies (e.g., Mollie Beattie, U.S. Fish and Wildlife Service; Jack Ward Thomas, U.S. Forest Service; Roger Griffis and Katharine Kimball, National Oceanic and Atmospheric Administration).

19. "Most people will not let the federal or state government on their property. They've got it posted. And the reason being is they're afraid that if [government officials] find something, this person from Fish and Wildlife, [for example,] finds something on their land that's a violation, that's going to be trouble for that farmer or property owner" (personal interview, 1999).

20. See also Weber 1998.

21. When stream flows drop in summer, water levels must be raised in order to deliver water to irrigation ditches. Farmers use heavy equipment to "push up" gravel and create dams, thus causing increased erosion and sediment flows (both bad for fisheries health), and creating barriers to passage (Applegate River Watershed Council, 1998).

22. Culverts can be too long, too high (i.e., fish needing to swim upstream cannot reach the water flowing through the culvert), damaged, or too steep (Applegate River Watershed Council, 1998).

23. When irrigation ditches cross streams, they sometimes capture and divert existing natural water flows. Depending on the season, the ditches can prevent spawning migrations, prevent juvenile salmon from out-migrating to the ocean, or both (Applegate River Watershed Council, 1998).

24. For example, in the Applegate community's vision statement, a high priority is given to making sure "the rural nature of the community is preserved with a good land use control program that promotes agriculture and low population density" (Priester and Moseley 1997, 15). Also see Applegate Partnership, 1997.

Chapter 5

1. The HFWC keeps a record of attendance for each meeting. Records include the name, organizational affiliation, and address of each individual participant. A total of 1,999 "visits" were made to the 46 meetings, making the average attendance for meetings 43.46 people.

2. Because the Targhee National Forest is federal land, control rests solely with federal authorities, in this case the USFS. As a result, there is no requirement to include local and state representatives in the decision process. Yet what the analysis of accountability suggests is that there is no absolute presumption in favor of broad-based accountability simply because a decision supports a national law. Some national laws might favor a broad set of interests; others are zero sum in character, like the ESA. The important point from the perspective of GREM and its critics is that the promise of staying true to existing mandates is matched by performance.

3. Noxious weeds are a policy problem because they can degrade wildlife habitat, choke streams and waterways, crowd out beneficial native plants, create fire hazards in forests and rangelands, and poison and injure livestock and humans. Not only does the problem "cost [the State of Idaho] millions of dollars," "the spread of noxious weeds may signal the decline of an entire ecological watershed" and "severely impact the beauty and biodiversity of natural areas. . . . Noxious weed species spare no segment of society—rancher, fisherman, and biker alike—and when unmanaged they spread rapidly, unceasingly, and silently" (Callihan and Miller 1997, v).

4. NRCS engineers ultimately offered their own alternative design. The question remains whether the agency would have done this on its own in the absence of the competing bioengineering design.

5. There were fifteen members of the Water Quality subcommittee as of June 2000. A broad range of perspectives were represented. Four members came from four different state agencies. Four federal participants also came from four different agencies. Two were independent scientists/consultants. Three worked with local Soils Conservation Districts and the final two came from one environmental organization (HFF) and one irrigation/farming organization (FMID), respectively (HFWC meeting minutes, June 20, 2000).

6. Over the years significant parts of the stream flow have been rerouted to irrigation canals due to broken and degraded water diversion structures.

7. Through a successful HFWC-initiated EPA Section 319 grant.

8. There are many public and private stakeholders from all levels in the Sheridan Creek project. They include the USFS, Idaho Fish and Game, Idaho Parks and Recreation, NRCS, ranchers/ private landowners, the HFF, and the FMID.

9. For the more general theme of the Bureau's poor environmental record, see Berkman and Vicusi 1973; Clarke and McCool 1996, 129–153; Liroff 1976; Reisner 1986; Wahl 1989.

10. Potential wording of such a section that has support among some HFWC

participants goes as follows: "Opportunities for inclusive stakeholder collaboration in the management of the water resources of the Henry's Fork Basin, as set forth in HCR 52, passed by the 52nd Idaho legislature in 1994, which supports the Council's 'innovative, consensus building process for sound watershed management' involving 'citizens, scientists and governmental agency representatives who reside, recreate, make a living, and/or have legal responsibilities in the Basin' " (Section 3[C] 1 [k] of Bruce Driver's proposed title transfer legislation, dated May 11, 1999).

11. Bruce Driver leads the Land and Water Conservation Fund of the Rockies. It represents a consortium of environmental groups including the Sierra Club, the Greater Yellowstone Coalition, and Idaho Rivers United, among others.

Chapter 6

1. Testimony of Dan'l Markham, executive director of the Willapa Alliance (U.S. Senate, 1997, 50–51).

2. See Washington State Department of Fish and Wildlife, 2000—a report that brings together 500 scientific studies and decades of research to document the vital role salmon play in the overall health of Northwestern U.S. ecosystems. Among other things, the report finds that 137 species depend on salmon, both alive and dead (as biomass), as a significant part of their diet. In fact, National Marine Fisheries Service biologist Robert Bilby "found that 40 to 50 percent of the stomach contents of young salmon and steelhead could be traced to salmon carcasses" (Hunt 2000, 2).

3. "Prioritization is based upon a fundamental philosophy which asserts that scarce resources should first be spent on the areas with the best opportunity for success and large returns. This approach lends [itself] to a strategy that points conservation and restoration efforts toward watersheds that are in less than healthy conditions, but healthy enough to be resilient and respond more quickly to efforts to improve their conditions" (Willapa Alliance, 1996c, 1).

4. A key part of this problem comes from the use of gillnets. Gillnets, because they catch fish by the gills, kill almost every fish hauled aboard fishing boats. Therefore, there is no chance to separate and save the wild fish except in a few cases.

5. Thus, "with the exception of five years since 1960 this species has dropped to record lows in the last three decades" (Willapa Alliance, 1998d, 29, 47–48).

6. They are called *dog salmon* because Native Americans used them in the past as a primary food source for their dogs.

7. See Willapa Alliance, 1998d, 31, chart 2-12. The 1998 WISC Report (Willopa Alliance, 1998d) also shows that chum commanded less than 50 cents per pound in twenty-two of the twenty-five years from 1970 to 1994 (31, chart 2-12).

8. Marston (1997a, 7) estimates that there are "up to 1 billion pounds of mud shrimp in the bay." Brett Dunbald, a scientist with the Washington Department of Fish and Wildlife, argues that "one can throw up a bunch of reasons [for the

proliferation of burrowing shrimp in the bay] and you can't prove any of them. There is the correlation with El Niño events, declines in sturgeon and salmon which feed on the larvae, [and] sediment from streams and logging" (Hunt 1995a, 7). There is a consensus, however, that while the lack of chum may not be the sole cause of the explosion in shrimp, it is one of the most important factors (see Hunt 1995a; Marston 1997a).

9. The alliance's 1998 *Willapa Indicators for a Sustainable Environment* report notes that Willapa has five miles of road per square mile with "logging roads . . . the most common type" (Willapa Alliance, 1998d, 31).

10. Funding is provided to agencies managing spartina eradication efforts as well as in the form of cost sharing with private-sector eradication efforts by oyster growers and crabbers.

11. Economic opportunities are measured by unemployment rates, the rates of in-migration and out-migration, the number of businesses, and new construction activity (see Willapa Alliance, 1996g, 14–15).

12. The 1996 WISC report (Willapa Alliance, 1996g) argues that "the lack of diversity in a community's industrial base sets the stage for a cycle of economic booms and busts . . . [which are] very destructive to the fabric of the community. . . . [By contrast, an increase in diversity is likely] "to strengthen . . . the sustainability and long-term security of the economy" (Willapa Alliance, 1996g, 16).

13. The equity component tries to assess the ability of the area economy to meet basic needs of citizens such as food and shelter by measuring poverty (How many live below the poverty line?), the extent of affordable housing, and the distribution of income (see Willapa Alliance, 1996g, 18–20).

14. The Alliance considers this an important indicator because citizens engaged in lifelong learning, defined as the number of people with high school diplomas and college degrees, are more likely to be self-motivated, independent, and possessing a general interest in learning and growth. Such citizens also will be better prepared to participate in the new high-skill economy (Willapa Alliance, 1996g, 20–21).

15. See Allen 1992; Blumenthal 1997; Hunt 1995b, 1995c; personal interviews, 1998.

16. Sources of money for the revolving loan fund are foundation grants and program-related investments, including a loan from Weyerhaeuser for $450,000 (Blumenthal 1997; Hunt 1995c, 12). In the beginning, ShoreTrust also had access to $5.3 million through a program called *EcoDeposits*. EcoDeposits "is a nationally marketed campaign that offers FDIC-insured savings, checking and money market accounts paying market rates, with the understanding that the money put into these funds will be invested in 'green' wholesale lending" (Hunt 1995c, 12). The longer-term goal of the ShoreTrust venture was to establish a new "commercial bank to complement and enhance the higher-risk credit available from Shore-Trust Trading" (Hunt 1995c, 12). The commercial bank, Shorebank Pacific, opened in July 1997 with a $12.4 million capitalization, including $2 million from the Ford Foundation, $7.5 million invested by 750 investors from forty

states and ten countries, and the remaining EcoDeposits moncy (Blumenthal 1997).

17. The National Fish and Wildlife Foundation, Bullitt Foundation, Sequoia Foundation, Forest Foundation, Hugh and Jane Ferguson Foundation, Meyer Memorial Trust, and Weyerhaeuser Corporation Foundation (Willapa Alliance, 1996c, 2).

18. The Ocean Beach School District, Port of Willapa, and Pacific County Conservation District.

Chapter 7

1. See Hempel 1999, 47, box 2.1, for eleven different definitions and conceptual approaches to sustainability.

2. Lackey (1998, 9) goes as far as to argue that "the sustainability concept until now has been mainly a social philosophy of what we 'ought' to do, rather than a scientific 'what is' formulation."

3. Hempel (1999, 52) notes that "interest in community-oriented strategies has grown as something of an adjunct to sustainable development ideas in general. Critics who dismissed the term *sustainable development* as oxymoronic welcomed the opportunity to replace the word *development* with *community*."

4. Also see Willapa Alliance, 1996g, 5; Applegate Partnership, 1996b, 1.

5. The attempts to determine just what sustainability is and how it will be achieved tend to be top-down and expert-led, with limited public input, a command-and-control orientation, and indicators "developed by scientists for scientists" (Bell and Morse 1999, 48; Hannum-Buffington 2000; Kates et al. 2001; Little 1997; Scruggs and Associates, 1996).

6. See Mazmanian and Kraft 1999a, 1999b; Hempel 1998; Nijkamp and Perrels 1994. The move toward ecosystem management, more generally, also "require[s] a much more active role for citizens than was true of past resource planning and management efforts" (Cortner and Moote 1999, 61).

7. See also Daniels 2001; Kates et al. 2001; Mazmanian and Kraft 1999a; Hempel 1999.

8. Other sources include Marston 1997c, 1; personal interviews, 1998 and 1999. According to Daniels (2001), such learning is essential because the sustainable-communities approach turns traditional government programs on their heads. Instead of government experts "administer[ing] programs *for* citizens, once sustainable community programs are created . . . citizens . . . administer their own programs *in collaboration with* . . . [government] officials" (xi; emphasis added).

9. Trade-offs among goals, of course, are still inevitable at the level of individual decisions. Yet the philosophy argues that every choice be made with an eye toward the overall health and maintenance of the larger system, defined as it is by the three major goals of environmental, economic, and community health.

10. For example, all participants are expected to endorse the "environment, economy, and community" mission and the holistic, integrated approach to natural resource management.

11. See Moseley 1999, 63; Priester and Kent 1997.

12. The Emergency Supplemental Appropriations Act (P.L. 104–119, 109 Stat. 240–247).

13. See sec. 2001 (a)(3) of the law.

14. See Lowe 1993; General Accounting Office, 1997. There have always been exceptions to the rule. In a 1992 memorandum, for example, timber planners in the Malheur National Forest were instructed to identify sales as salvage "as long as one board comes off that would qualify as salvage" (Boaz 1992, 2).

15. Challenges are required to be filed in local U.S. District Courts (where the lands in question were located) within fifteen days of the initial advertisement of the sale. "Further, courts are directed to generally render final decisions within 45 days, and are prohibited from granting restraining orders, preliminary injunctions, or injunctions pending appeal" (Gorte 1996, 2).

16. The Applegate Partnership maintains a mailing list of over 300 people that goes to participants as well as individuals and organizations outside of the Applegate region, including the Sierra Club, Wilderness Society, Nature Conservancy, and National Audubon Society.

17. A reader elaborates: "Every single issue in the beginning had a map of the Applegate and you'd open it up and there would be a theme, maybe it would just be a map. The next issue would be here's where fish are in the Applegate. Another one was here's stores in the Applegate, here's fire districts. And there would be articles about fire districts, or stores, or fish. And then you see yourself as part of the Applegate on that map too. Over time, I think what happened was that people began to see themselves as part of the Applegate community much more so than ever before" (personal interview, 1999).

18. Other topics have included, among others, (1) hatcheries and hatchery reform, (2) the Washington State Wild Salmonid Policy, the Governor's Salmon Plan, and the State Legislature's Salmon Recovery Plan, and (3) Water Productivity and Water Quality—Fact, Fiction, and Potential Impacts (Willapa Alliance and Pacific County Economic Development Council, 1997).

19. This is done in cooperation with Interrain Pacific.

20. A few recent examples of conferences attended by people involved in the HFWC are the Seventh Biennial Watershed Management Council conference (Boise, ID, October 1998), the Visiting Lecturer Series of the Natural Resources and Environmental Policy Center, Utah State University (Logan, UT, October 1998), the Trout Unlimited national convention (August 1999), and the American Society of Civil Engineers watershed resources consortium (Seattle, August 1999). For their part, Applegate Partnership participants have attended meetings like the Lead Partnership Group's roundtable on *Communities of Place, Partnerships, and Forest Health* (Blairsden, CA, 1995) and the national consortium for Community-Based Collaborative Conservation (Tucson, AZ, October 1999).

According to one participant, "[Jack] Shipley and I have attended workshops in South Dakota, in Portland, Oregon, and in other places too numerous to count. We go and tell the story of the Partnership and what we're trying to do" (personal interview, 1999).

21. Jack Shipley, the source of the remarks just quoted, also attended a Sierra Club board of directors meeting that same year and invited them to have a representative sit on the Partnership board. They said no (personal interview, 1999).

22. The estimate is based on a sampling of 205 meetings over a six-year period. Hughes attended 163 of the meetings.

23. Willapa Alliance 1995, 6; Willapa Alliance, 1996e, 4; personal interview, 1998. In one year alone, the Willapa Alliance attracted 162 volunteers who together "logged some 2,000 hours of work" (Willapa Alliance, 1996e, 5).

24. Participants are not naive. They do not expect all landowners to adopt more sustainable practices as a result of incentives. They know that some will choose to remain ignorant, while others will simply refuse to change established practices no matter how much information they are exposed to. In these cases, top-down regulatory action is probably needed to achieve the public goal. However, a cooperative approach sensitive to incentives as a motivation for action recognizes that there is a category of citizens who might normally resist regulators' attempts to impose solutions from the top down, but who will willingly accept and adopt the same goals if sustainability is framed as a "good," or profitable business practice. The logic is similar to the idea of responsive regulation and the ability of policy administrators to distinguish between good and bad apples (see Ayres and Braithwaite 1992).

25. Bill Bradbury, former president of the Oregon Senate and now director of For the Sake of the Salmon in Oregon, explains that "what you can get through collaboration is a lot more than the minimum effort [so typical of traditional, top-down regulation] from landowners who recognize they've got a stake in sustaining ecological systems where they live. . . . As part of Oregon's [voluntary, collaborative] salmon recovery effort, for instance, timber companies have pledged to spend millions . . . to stabilize old logging roads on their own private forestlands that now shed sediment into salmon spawning streams" (as quoted in Arrandale 1997, 66).

26. "Hub" grazing involves portable, New Zealand–style fencing (electric, single wire) in combination with corral fencing and numerous gates that restrict cattle to a particular section of range and channel them toward the new watering troughs (the hub). After foraging on the same range for an appropriate amount of time, the portable fencing is moved in clockwise fashion around the watering trough "hub" to encompass a new section of range, while still maintaining the same watering hub.

27. See also the case of the Feather River Alliance, an example of GREM in California. Little (1997, 7, 11) reports that "instead of relying [entirely] on outside specialists, alliance members are using local expertise and knowledge of how streams have changed to help determine how they can be restored. . . . The initiative taken by local community groups to solve resource problems represents

a sea change in the process for making land management decisions. In place of fixes conceived and developed by [distant] urban bureaucrats with few ties to a particular place, the people who live in those places are insisting on being part of the solution. . . . The Feather River [Alliance is] part of a movement that is reordering the nation's top-down society to include rural communities in the responsibility for managing natural resources."

28. This is a key tenet of the emerging field of sustainability science (see Kates et al. 2001, but also Hanna, Folke, and Maler 1997b). MacDonnell (1999) makes the same point, while also providing a good account of how the current laws and institutions governing Western U.S. water resources are poorly matched with ecological processes.

29. See Bell and Morse 1999; Heinen 1994; Hempel 1999; Mazmanian and Kraft 1999a. The Bellagio Principles for sustainability (deriving from a meeting in Bellagio, Italy, funded by the Rockefeller Foundation) and Chapter 40 of the 1992 Rio Earth Summit Agenda 21 document also call for the development of indicators based on standardized measurement as a necessary step in the effort to achieve sustainability.

31. See Bell and Morse 1999, especially chap. 1.

Chapter 8

1. See the corporatism literature (e.g., Lehmbruch and Schmitter 1982).

2. See Cortner and Moote 1999, 60; Moe 1994. Behn (2001) also wrestles with this dilemma, and raises the possibility of mutual, collective responsibility, yet does not offer any empirical evidence where such a system is in operation.

3. The final sentences of this paragraph are a close paraphrase of Behn 1999, 132. Critics of the new public management paradigm include Fredrickson (1992), Moe (1994), and Moe and Gilmour (1995), among others.

4. Behn (1999, 143) calls this the "most troubling question raised by the new public management paradigm."

5. On the last point, it is inevitable "due to the fragmentation of authority inherent in Congress's decentralized structure" (West and Cooper 1989–90, 588; see also Fiorina 1989).

6. Mosher (1982) argues that the professionalization of bureaucracies, when combined with the principles of administration, helps experts capture public agencies. Professional capture, however, does not always equate with either efficiency or accountability. Evaluation systems grounded in written rules are apt to discourage innovation and creative thinking and to encourage rigid adherence to the rules so as to ensure career advancement, thus limiting the problem-solving capacity of the agency in question. Professional domination of an agency also can lead to goal displacement, where the rules and procedures become more important than the overarching *public* goals of the organization itself. Moreover, trained incapacity—a narrow outlook on problem solving and possible alternatives—is more likely in an agency staffed and controlled by a single profession.

The lack of alternative "expert" input over time can do serious damage to a public agency's ability to respond to a changing political context and changing sets of policy problems. Finally, as Herbert Finer (1941) tells us, the professional standards guiding agency decision making in a world of delegated authority will not necessarily match public needs as defined by presidents, Congress, or the courts.

7. For problems of inefficiency and goal achievement, see, for example, Light 1993; Rabe 1994; Weber 1998, 70–104. For how participation "rights" are largely limited to experts and advocates from organized interests, see Rosenbaum 1976; Sirianni and Friedland 1995a, 2001. For how centralization of government authority, growth of procedural encumbrances, and expanded use of coercion are key contributors to growing citizen disenchantment with and concern over the legitimacy of government decisions, see Newland 1996; Sandel 1996. For example, the allegiance to social equity and the expansion of individuals' rights advocated by the public interest–egalitarian model limits the potential for deliberation and compromise often crucial for resolving competing political demands. Couching policy issues in the language of absolute rights obviates the need for deliberation, since such rights are nonnegotiable. It contributes to the perception of a government that is unaccountable by virtue of having already decided many important policy decisions pertaining to the collective welfare (i.e., removing an issue from the realm of politics). It also dampens the incentive for communities to cooperate for the collective good and, in some cases, removes the ability of communities to decide policy for themselves (see Glendon 1991; Sandel 1996).

8. Some are likely to insist that any imbalances, either here or in terms of diversity in GREM proceedings, be offset with the required presence of national *public/environmental* interest group representatives. The suggestion is usually made under the assumption that these groups somehow represent the general public better than local interests. But see the discussion in Sirianni and Friedland 2001, 11, for concerns that this may not always be true, especially when national public interest groups organize for action at the local level. See also Dahl 1994.

9. The assumption is that participants select appropriate indicators. A poor choice of an indicator, of course, lessens accuracy and hampers assessment.

10. See also Selznick (1949), where he talks about infusing agencies with a particular value important to the overall mission.

11. A virtuous citizenry does not seem a necessary precondition for broad-based accountability *at the beginning of the effort*. Rather, it seems that the kinds of virtue important to broad-based accountability—law-abidingness, civility, the capacity to delay gratification, independence, entrepreneurialness, tolerance and inclusiveness, and honesty and integrity—can be produced by extensive exposure to the kinds of cooperative practices found in GREM. See chapter 3, but also Kemmis 1990, 77–81; Galston 1995, 38; Glendon 1995; Sandel 1996.

12. See Ingraham and Kneedler 2000.

13. Rotation between management units does not have to stop, it just needs to be done less frequently. Rotation transfers can still be used as a tool for gaining additional expertise.

14. The ultimate success of a regular reporting system is directly connected to the transparency factor. Without transparency it will be difficult to access the kinds of information necessary for outcome assessment.

15. Seven of the ten diversion structures slated for reconstruction are on his property.

16. Williams and Matheny (1995, 170) make the same argument when discussing the communitarian discourse.

17. See Dryzek 1987; Kemmis 1990; Majone 1990; Reich 1990a, 1990b; Scott 1998; Torgerson 1998. This position accepts the demise of the politics-administration dichotomy. As government has grown larger over the past seventy years and greater discretion has been granted to bureaucrats, more bureaucratic decisions are *policymaking* exercises, rather than neutral implementation of law. For but a few of the treatises making this clear, see Long [1947] 1986; Allison 1971; Knott and Miller 1987; Behn 1999. This is not to say that administrators are willing to accept such a conclusion. In part 1 of a five-part *Washington Post* series exposing the depth to which politics influences the U.S. Army Corps of Engineers, written responses by General Joe Ballard, the agency's recently retired chief engineer, explain that "the Corps [is] a model of public service . . . an apolitical military organization, simply following orders produced by a democratic process" (Grunwald 2000a, A1).

18. See Scott 1998. He explores the problems associated with a number of case studies, including the Soviet collectivization of agriculture; Tanzanian "villagization," urban planning (using Brasilia, the capital city of Brazil, as the focal point), and the ecological problems stemming from the scientific management of forests.

19. See, for example, Brick, Snow, and van de Wetering 2001; Cestero 1999; Kenney 1999, Lubell 2000; McKinney 2001; Rabe 1994; Weber 1998; Wondollek and Yaffee 2000. Paul Sabatier has also organized the Watershed Partnerships Project at the University of California–Davis for the express purpose of developing a better understanding of why collaboratives succeed or fail (see, for example, Leach and Pelkey 2000; Pelkey et al. 1999). There are also publications that do not focus specifically on environmental policy, including Bardach 1998, Chrislip and Larson 1994, Fisher and Ury 1981, Gray 1985, and Susskind and Cruickshank 1987, among others.

References

Allen, William. 1992. "Region Seeks to Protect What Provides Its Living, Bay Residents Want Development 'Without Fouling Our Nest.'" *St. Louis Post Dispatch* (May 12): 1A.

Allison, Graham T. 1971. *The Essence of Decision.* Boston: Little, Brown.

Altshuler, Alan, and William Parent. 1997. "Breaking Old Rules: Four Themes for the 21st Century." In *Achieving Excellence, Building Trust,* 11–14. Annual Report of the Ford Foundation on the *Innovations in American Government Awards Program.* New York, NY: The Ford Foundation.

Amy, Douglas J. 1987. *The Politics of Environmental Mediation.* New York: Columbia University Press.

Applegate Partnership. 1995. *Stewardship in the Applegate Valley: Issues and Opportunities in Watershed Restoration.* Jacksonville, OR: Applegate Partnership.

Applegate Partnership. 1996a. *A Home for All of Us.* Jacksonville, OR: Applegate Partnership.

Applegate Partnership. 1996b. "An Open Letter to the Environmental Community [letter to the editor]." *High Country News.* Online at www.hcn.org/home_page/dir/email_letters.html; retrieved on January 20, 1998.

Applegate Partnership. 1997. "Stewardship Contracting." Meeting minutes (January 22).

Applegate Partnership. 1999. Meeting minutes (February 3). Applegate Partnership. 1993. Meeting minutes (July 7).

Applegate River Watershed Council. 1998. *Applegate River Watershed Council's Restoration Program.* Jacksonville, OR: Applegate River Watershed Council.

Arendt, Hannah. 1959. *The Human Condition.* Garden City, NJ: Doubleday.

Arrandale, Tom. 1997. "Conservation by Consensus." *Governing,* (July): 66.

Ayres, Ian, and John Braithwaite. 1992. *Responsive Regulation: Transcending the Deregulation Debate.* New York: Oxford University Press.

Bahro, Rudolf. 1986. *Building the Green Movement.* Philadelphia: New Society Books.

Barber, Benjamin R. 1984. *Strong Democracy: Participatory Politics for a New Age.* Berkeley: University of California Press.

Bardach, Eugene. 1998. *Getting Agencies to Work Together: The Practice and Theory of Managerial Craftsmanship.* Washington, DC: Brookings Institution Press.

Bardach, Eugene, and Robert A. Kagan. 1982. *Going by the Book: The Problem of Regulatory Unreasonableness.* Philadelphia: Temple University Press.

Bardach, Eugene, and Cara Lesser. 1996. "Accountability in Human Services Collaboratives—For What? and To Whom?" *Journal of Public Administration Research and Theory,* 6 (April): 197–224.

Barker, Robert. 1995. "Old Foes Join to Protect Henry's Fork." *Post Register* (September 26): B1.

Baumgartner, Frank R., and Bryan D. Jones. 1993. *Agendas and Instability in American Politics.* Chicago: University of Chicago Press.

Bayley, David. H. 1994. *Police for the Future.* New York: Oxford University Press.

Behn, Robert D. 1999. "The New Public-Management Paradigm and the Search for Democratic Accountability." *International Public Management Journal,* 1 (2): 131–165.

Behn, Robert D. 2001. *Rethinking Democratic Accountability.* Washington, DC: Brookings Institution Press.

Beierle, Thomas C., and Jerry Cayford. 2001. Democracy in Partice: *Public Participation in Environmental Decisions.* Washington, DC: Resources For the Future Press.

Bell, Simon, and Stephen Morse. 1999. *Sustainability Indicators: Measuring the Immeasurable?* London: Earthscan.

Benjamin, Lynn. 2000. "An Investigation of the Groundwater Hydrology of the Henry's Fork Springs." *Intermountain Journal of Sciences,* 6 (3) (September): 119–142.

Benjamin, Lynn, and Robert W. Van Kirk. 1999. "Assessing Instream Flows and Reservoir Operations on an Eastern Idaho River." *Journal of the American Water Resources Association,* 35: 899–909.

Berger, Peter L., and Richard John Neuhaus. 1996. *To Empower People: From State to Civil Society* (edited by Michael Novak). 2nd ed. Washington, DC: The AEI Press.

Berkman, Richard L., and W. Kip Vicusi. 1973. *Damning the West.* New York: Grossman.

Bernstein, Marver. 1955. *Regulating Business by Independent Commission.* Princeton, NJ: Princeton University Press.

Bingham, Gail. 1986. *Resolving Environmental Disputes: A Decade of Experience.* Washington, DC: Conservation Foundation.

Blumenthal, Les. 1997. "Fledgling Lender Banks on Its Ability to Help Ecosystem, Economy." *News Tribune* (April 27): B13.

Boaz, B. 1992. Message to staff. USDA Forest Service, Malheur National Forest, Oregon (December 17).

Born, Steve, and Kenneth Genskow. 1999. *Exploring the Watershed Approach: Critical Dimensions of State-Local Partnerships.* Portland, OR: River Network.

Box, Richard C. 1998. *Citizen Governance: Leading American Communities into the 21st Century.* Thousand Oaks, CA: SAGE.

Braithwaite, John. 1998. "Institutionalizing Distrust, Enculturating Trust." In V. Braithwaite and M. Levi, eds., *Trust and Governance,* 343–375. New York: Russell Sage Foundation.

Braxton, Jane. 1995. "The Quincy Library Group." *American Forests,* 101 (January-February): 21–23.

Brick, Philip, Donald Snow, and Sarah van de Wetering, eds. 2001. *Across the Great Divide: Explorations in Collaborative Conservation and the American West.* Washington, DC: Island Press.

Broder, David. 1994. "The Citizen's Movement." *Washington Post* (November 15): A20.

Bromley, Daniel W., ed. 1992. *Making the Commons Work: Theory, Practice, and Policy.* San Francisco: ICS Press.

Brower, David, with Steve Chappel. 1995. *Let the Mountains Talk, Let the Rivers Run.* San Francisco: HarperCollins West.

Brown, Janice M. 1996a. "Buffalo and Sheridan Projects Illustrate the Challenges of the Stewardship Program." *Henry's Fork Foundation Quarterly Newsletter,* (Fall): 5–7.

Brown, Janice M. 1996b. On the Henry's Fork: Building a Trust Account. *Outdoor Idaho* website (November 5): idptv.state.id.us/outdoors/shows/henry/art.html.

Brown, Janice M. 1999. Public Remarks. Eastern Idaho Watershed Conference, Pocatello (October 21).

Burke, John P. 1986. *Bureaucratic Responsibility.* Baltimore: Johns Hopkins University Press.

Callihan, Robert H., and Timothy W. Miller. 1997. *Idaho's Noxious Weeds: The Cancer of Our Land.* Boise: Noxious Weed Advisory Council, Idaho Department of Agriculture (July).

Caro, Robert A. 1975. *The Power Broker: Robert Moses and the Fall of New York.* New York: Vintage Books.

Center for the New West. 1994. *The Lone Eagle Reading File.* Denver: Center for the New West.

Cestero, Barb. 1999. *Beyond the Hundredth Meeting: A Field Guide to Collaborative Conservation on the West's Public Lands.* Tucson, AZ: Sonoran Institute.

Chertow, Marian R., and Daniel C. Esty. 1997. *Thinking Ecologically: The Next Generation of Environmental Policy*. New Haven, CT: Yale University Press.

Chinook Observer. 1996. "A Good Start: Weekend Conference Made Progress in Uniting Pacific County Citizens." Online at www.ecotrust.org/wisconf.htm (March 26); retrieved on January 25, 1997.

Chrislip, David D., and Carl E. Larson. 1994. *Collaborative Leadership: How Citizens and Civic Leaders Can Make a Difference*. San Francisco: Jossey-Bass.

Christensen, Jon. 1996. "Everyone Helps a California Forest—Except the Forest Service." *High Country News*, 28 (9) (May 13): 8–9.

Christensen, Kelly. 2000. "New Weapons in the War on the Bay." *Tidepool*. Online at www.tidepool.org/ctc/spartina.cfm (June 20); retrieved on October 10, 2000.

Clarke, Jeanne, and Daniel McCool. 1996. *Staking Out the Terrain: Power and Performance among Natural Resource Agencies*. 2nd ed. Albany, NY: SUNY Press.

Cleveland, John. 1997. *Steady Progress: A Review of Ecotrust's Community Strategy*. Lansing, MI On Purpose Associates (October 15).

Cockburn, Alexander. 1997. "Scarring Forests in the Name of Conservation." *Arizona Republic* (July 24): B7.

Coggins, George C. 1998. "Of Californicators, Quislings and Crazies: Some Perils of Devolved Collaboration." *Chronicle of Community*, 2 (2) (Winter): 27–33.

Community-Based Collaboratives Workshop. 1999. Tucson, AZ. Sponsored by the Udall Center for Studies in Public Policy, University of Arizona, and the Institute for Environmental Mediation, University of Virginia (October 8–9).

Cortner, Hanna J., and Margaret A. Moote. 1999. *The Politics of Ecosystem Management*. Washington, DC: Island Press.

Costanza, Robert, and Carl Folke. 1997. "The Structure and Function of Ecological Systems in Relation to Property-Rights Regimes." In Susan Hanna, Carl Folke, and Karl-Goran Maler, eds. *Rights to Nature: Ecological, Economic, Cultural, and Political Principles of Institutions for the Environment*, 13–34. Washington, DC: Island Press.

Culhane, Paul J. 1981. *Public Lands Politics: Interest Group Influence on the Forest Service and the Bureau of Land Management*. Baltimore: The Johns Hopkins University Press.

Cushman, John H., Jr. 1999. "Audit Faults Forest Service on Logging Damage in U.S. Forests." *New York Times* (February 5): A5.

Dagget, Dan. 1995. *Beyond the Rangeland Conflict: Toward a West That Works*. Layton, UT: Gibbs Smith.

Dahl, Robert. 1994. *The New American Political (Dis)Order: An Essay*. Berkeley, CA: Institute of Governmental Studies.

Daniels, Mark R., ed. 2001. *Creating Sustainable Community Programs: Examples of Collaborative Public Administration*. Westport, CT: Praeger.

Davies, J. Clarence. 1990. "The United States: Experiment and Fragmentation." In Nigel Haigh and Frances Irwin, eds., *Integrated Pollution Control in Europe and North America,* 51–66. Washington, DC: Conservation Foundation.

Davis, Charles, and M. Dawn King. 2000. "The Quincy Library Group and Collaborative Planning within U.S. National Forests." Online at www.qlg.org/pub/Perspectives/daviskingcasestudy.htm (May); retrieved on August 20, 2000.

Dawes, Robyn M., Alphons J. C. van de Kragt, and John M. Orbell. 1990. "Cooperation for the Benefit of Us—Not Me, or My Conscience." In Jane Mansbridge, ed., *Beyond Self-Interest,* 97–110. Chicago: University of Chicago Press.

DeLeon, Linda, and Robert B. Denhardt. 2000. "The Political Theory of Reinvention." *Public Administration Review,* 60 (2) (March-April): 89–97.

Delli Carpini, Michael X., and Scott Keeter. 1996. *What Americans Know about Politics and Why It Matters.* New Haven, CT: Yale University Press.

Dionne, E. J., Jr., ed. 1998. *Community Works: The Revival of Civil Society in America.* Washington, DC: Brookings Institution.

Downs, Martha. 2000. "Researchers Reach Out to 'Stakeholders' in Studies." *Chronicle of Higher Education* (August 18) (www.chronicle.com).

Dryzek, John S. 1987. *Rational Ecology: Environment and Political Economy.* New York: Blackwell.

Dryzek, John S. 1997. *The Politics of the Earth: Environmental Discourses.* New York: Oxford University Press.

Durant, Robert F. 1998. "Agenda Setting, the 'Third Wave,' and the Administrative State." *Administration and Society,* 30 (3) (July): 211–247.

Durbin, Kathie. 1996. *Tree Huggers: Victory, Defeat, and Renewal in the Northwest Ancient Forest Campaign.* Seattle: The Mountaineers Press.

Durning, Alan T. 1996. *This Place on Earth: Home and the Practice of Permanence.* Seattle: Sasquatch Books.

Ecological Applications. 1996. *Special Issue: Perspectives on Ecosystem Management,* 6 (3) (August).

Ellickson, Robert. 1991. *Order without Law: How Neighbors Settle Disputes.* Cambridge, MA: Harvard University Press.

Ensminger, Jean. 1996. "Culture and Property Rights." In S. Hanna, C. Folke, and K. G. Maler, eds., *Rights to Nature: Ecological, Economic, Cultural, and Political Principles of Institutions for the Environment,* 179–204. Washington, DC: Island Press.

Etzioni, Amitai. 1996. *The New Golden Rule: Community and Morality in a Democratic Society.* New York: Basic Books.

Etzioni, Amitai. 1998. *The Essential Communitarian Reader.* New York: Rowman and Littlefield.

Fesler, James. W., and Donald F. Kettl. 1991. *The Politics of the Administrative Process.* Chatham, NJ: Chatham House.

Finer, Herman. 1941. "Administrative Responsibility in Democratic Government." *Public Administration Review,* 1: 335–350.

Fiorina, Morris P. 1989. *Congress: Keystone of the Washington Establishment.* 2nd ed. New Haven, CT: Yale University Press.

Fisher, Roger, and William Ury. 1981. *Getting to Yes.* Boston: Houghton Mifflin.

Fredrickson, George. 1992. "Painting Bull's Eyes around Bullet Holes." *Governing,* 6 (December): 60–65.

Frederickson, George. 1996. "Comparing the Reinventing Government Movement with the New Public Administration." *Public Administration Review,* 56 (May-June), 263–269.

Fremont County Herald-Chronicle. 1995. "Council Ready to Discuss Frogs, Dam Title Transfers at Meeting." (November 16): 3A.

Fremont County Herald-Chronicle. 1995. "River Temperature Crisis Averted by Cooperation." (July 27): 6A.

Friedrich, Carl J. 1940. "Public Policy and the Nature of Administrative Responsibility." *Public Policy,* 1: 1–20.

Fukuyama, Francis. 1995. *Trust: The Social Virtues and the Creation of Prosperity.* New York: Free Press.

Galanter, Marc. 1974 "Why the 'Haves' Come Out Ahead: Speculations on the Limits of Legal Change." *Law and Society Review* 9: 95–160.

Galston, William A. 1995. "Liberal Virtues and the Formation of Civic Character." In Mary Ann Glendon and David Blankenhorn, eds., *Seedbeds of Virtue: Sources of Competence, Character, and Citizenship in American Society,* 35–60. Lanham, MD: Madison Books.

Geddes, Pete. 1998. "Economy and Ecology in the Next West." *Journal of Forestry,* 96 (3): 46–59.

General Accounting Office. 1994. *Ecosystem Management: Additional Actions Needed to Adequately Test a Promising Approach.* (RCED-94-111). Washington, DC: General Accounting Office.

General Accounting Office. 1997. *Emergency Salvage Sale Program.* (RCED-97-53). Washington, DC: General Accounting Office.

General Accounting Office. 1998. "Forest Service: Lack of Financial Accountability Has Resulted in Inefficiency and Waste." Statement of Barry T. Hill, Associate Director, Energy, Resources, and Science Issues. (March 26) (T-RCED/AIMD-98-135). Washington, DC: General Accounting office.

Getches, David H. 1995. "Foreword." In Dan Dagget, ed., *Beyond the Rangeland Conflict: Toward a West That Works,* vii–viii. Layton, UT: Gibbs Smith.

Gilmour, Robert S., and Laura S. Jensen. 1998. "Reinventing Government Accountability: Public Functions, Privatization, and the Meaning of State Action." *Public Administration Review,* 58 (3) (May-June): 247–258.

Glendon, Mary Ann. 1991. *Rights Talk: The Impoverishment of Political Discourse.* New York: Free Press.

Glendon, Mary Ann. 1995. "Introduction: Forgotten Questions." In Mary Ann Glendon and David Blankenhorn, eds., *Seedbeds of Virtue: Sources of Competence, Character, and Citizenship in American Society*, 1–16. Lanham, MD: Madison Books.

Glendon, Mary Ann. 1997. *The Current State of the "Civil Society" Debate*. New York: Institute for American Values.

Glendon, Mary Ann, and David Blankenhorn, eds. 1995. *Seedbeds of Virtue: Sources of Competence, Character, and Citizenship in American Society*. Lanham, MD: Madison Books.

Goldstein, Bruce E. 1999. "Combining Science and Place-Based Knowledge: Pragmatic and Visionary Approaches to Bioregional Understanding." In Michael V. McGinnis, ed., *Bioregionalism*, 176–195. New York: Routledge.

Gormley, William T., Jr. 1989. *Taming the Bureaucracy: Muscles, Prayers, and Other Strategies*. Princeton, NJ: Princeton University Press.

Gormley, William T., Jr. 1995. *Everybody's Children: Child Care as a Public Problem*. Washington, DC: Brookings Institution.

Gorte, Russ W. 1996. "The Salvage Timber Sale Rider: Overview and Policy Issues." *Congressional Research Service* (June 24) (96–569 ENR).

Gottlieb, Robert, and Margaret FitzSimmons. 1991. *Thirst for Growth: Water Agencies as Hidden Government in California*. Tucson: University of Arizona Press.

Grand Canyon Trust. 1999. "A Declaration of Principles for Responsible Forest Restoration." Draft proposal (May). Flagstaff, AZ: Grand Canyon Trust.

Gray, Barbara. 1985. "Conditions Facilitating Interorganizational Collaboration." *Human Relations*, 38: 910–926.

Gregory, John. 1997. *Upper Henry's Fork Habitat Assessment, Headwaters to Island Park Dam, Summer 1996*. Project completion report for Henry's Fork Foundation, Ashton, ID. Mackay, ID: Gregory Aquatics.

Grunwald, Michael. 2000a. "Engineers of Power Inside the Army Corps." *Washington Post* (September 10): A1.

Grunwald, Michael. 2000b. "A Race to the Bottom." *Washington Post* (September 12): A1.

Grunwald, Michael. 2000c. "Reluctant Regulator on Alaska's North Slope." *Washington Post* (September 13): A1.

Grunwald, Michael. 2000d. "Working to Please Hill Commanders." *Washington Post* (September 11): A1.

Gulick, Luther. [1937]. 1997. "Notes on the Theory of Organization." In J. M. Shafritz and A. C. Hyde, eds., *Classics of Public Administration*, 81–89. 4th ed. Fort Worth, TX: Harcourt Brace.

Haeuber, Richard. 1996. "Setting the Environmental Policy Agenda: The Case of Ecosystem Management." *Natural Resources Journal*, 36 (1) (Winter): 1–28.

Hall, Bob, and Mary Lee Kerr. 1992. *1991–1992 Green Index: A State-by-State Guide to the Nation's Environmental Health*. Washington, DC: Island Press.

Hammond, Thomas H., and Jack H. Knott. 1996. "Who Controls the Bureaucracy? Presidential Power, Congressional Dominance, Legal Constraints, and Bureaucratic Autonomy in a Model of Multi-Institutional Policy-Making." *Journal of Law, Economics, and Organization,* 12(1): 121–168.

Hanna, Susan, Carl Folke, and Karl-Goran Maler. 1997a. "Property Rights and the Natural Environment." In Susan Hanna, Carl Folke, and Karl-Goran Maler, eds., *Rights to Nature: Ecological, Economic, Cultural, and Political Principles of Institutions for the Environment, 1–10.* Washington, DC: Island Press.

Hanna, Susan, Carl Folke, and Karl-Goran Maler, eds. 1997b. *Rights to Nature: Ecological, Economic, Cultural, and Political Principles of Institutions for the Environment.* Washington, DC: Island Press.

Hannum-Buffington, Christina L. 2000. *Developing a Community Indicators Project in the Applegate Watershed.* Unpublished master's thesis, Washington State University.

Harris, Charles, Greg Brown, and Bill McLaughlin. 1996. *Rural Communities in the Inland Northwest: An Assessment of the Past and Present.* Moscow, ID: College of Forestry, Wildlife and Range Sciences, University of Idaho.

Hays, Samuel P. 1959. *Conservation and the Gospel of Efficiency: The Progressive Conservation Movement, 1890–1920.* Cambridge, MA: Harvard University Press.

Heclo, Hugh. 1978. "Issue Networks and the Executive Establishment." In A. King, ed., *The New American Political System,* 88–120. Washington, DC: American Enterprise Institute.

Heinen, J. T. 1994. "Emerging, Diverging and Converging Paradigms on Sustainable Development." *International Journal of Sustainable Development and World Ecology,* 1: 22–33.

Hempel, Lamont C. 1992. "Earth Summit of Abyss?" Paper presented at the Global Forum, United Nations Conference on Environment and Development, Rio de Janeiro, Brazil (June).

Hempel, Lamont C. 1996. *Environmental Governance: The Global Challenge.* Washington, DC: Island Press.

Hempel, Lamont C. 1998. *Sustainable Communities: From Vision to Action.* Unpublished manuscript, Claremont Graduate School.

Hempel, Lamont C. 1999. "Conceptual and Analytical Challenges in Building Sustainable Communities." In D. A. Mazmanian and M. E. Kraft, eds., *Toward Sustainable Communities,* 43–74. Cambridge, MA: MIT Press.

Henry's Fork Watershed Council. 1996. *1996 Evaluation Report.* Seattle: Northwest Policy Center, University of Washington (November 25).

Henry's Fork Watershed Council. 1998. *A Community-Based Approach to Watershed Protection and Management.* Ashton, ID: Henry's Fork Watershed Council.

Henry's Fork Watershed Council. 1999. *Cutthroats on Thurmon Creek.* (October). Ashton, ID: Henry's Fork Watershed Council.

Hollander, C. V. 1995a. "Beauty, Purity of Willapa Bay Matches Strength of Its Community." In *Willapa: Banking on the Bay*, compiled by the *Daily Astorian* and the *Chinook Observer:* 2–3.

Hollander, C. V. 1995b. "Forks Creek Hatchery Fishes for the Future." In *Willapa: Banking on the Bay*, compiled by the *Daily Astorian* and the *Chinook Observer:* 14.

Hollander, C. V. 1995c. "Groups Cultivate Stake in Willapa Bay's Future." In *Willapa: Banking on the Bay*, compiled by the *Daily Astorian* and the *Chinook Observer:* 8–9.

Hollander, C. V. 1995d. "Oyster Industry Depends, Thrives in Bay's Clean Waters." In *Willapa: Banking on the Bay*, compiled by the *Daily Astorian* and the *Chinook Observer:* 7–8

Hollander, C. V. 1995e. "Social Turnabout Is Mission of Ecotrust Founder." In *Willapa: Banking on the Bay*, compiled by the *Daily Astorian* and the *Chinook Observer:* 15.

Hollander, C. V. 1995f. "Tribe Brings Concerns to Alliance Table." In *Willapa: Banking on the Bay*, compiled by the *Daily Astorian* and the *Chinook Observer:* 9.

Hollander, C. V. 1995g. "Willapa Is a First-of-Its-kind Wonder." In *Willapa: Banking on the Bay*, compiled by the *Daily Astorian* and the *Chinook Observer:* 15.

Hunt, Ed. 1995a. "Predators Thrive in the 'Pristine' Waters." In *Willapa: Banking on the Bay*, compiled by the *Daily Astorian* and the *Chinook Observer:* 7.

Hunt, Ed. 1995b. "ShoreTrust Brings Innovation to Funding on the Bay." In *Willapa: Banking on the Bay*, compiled by the *Daily Astorian* and the *Chinook Observer:* 12.

Hunt, Ed. 1995c. "Solutions for Bay Mix Capitalism, Conservation: Lenders Focus on 'Green' Loans." In *Willapa: Banking on the Bay*, compiled by the *Daily Astorian* and the *Chinook Observer:* 11–13.

Hunt, Ed. 2000. "New Report Finds 137 Species Depend on Salmon." *Tidepool*. Online at www.tidepool.org/salmon.depend.html; retrieved on October 12, 2000.

Inglehart, Ronald. 1997. "Postmaterialist Values and the Erosion of Institutional Authority." In J. Nye, P. Zelikow, and D. King, eds., *Why People Don't Trust Government*, 217–236. Cambridge, MA: Harvard University Press.

Ingraham, Patricia W., and Amy E. Kneedler. 2000. "Dissecting the Black Box: Toward a Model and Measures of Government Management Performance." In J. Brudney, L. O'Toole, and H. Rainey, eds., *Advancing Public Management*, 235–252. Washington, DC: Georgetown University Press.

Interagency Ecosystem Management Task Force. 1996. *The Ecosystem Approach: Healthy Ecosystems and Sustainable Economies, Volume III, Case Studies*. Washington, DC: U.S. Government Printing Office.

Iyengar, Shanto. 1991. *Is Anyone Responsible? How Television Frames Political Issues*. Chicago: University of Chicago Press.

Jackman, Robert A., and Ross A. Miller. 1998. "Social Capital and Politics." In Nelson Polsby, ed., *Annual Review of Political Science* 1, 47–73. Palo Alto, CA: Annual Reviews.

Jaeger, M., Robert Van Kirk, and T. Kellogg. 2000. "Status of Yellowstone Cutthroat Trout in the Henry's Fork Watershed." *Intermountain Journal of Sciences*, 6 (3) (September): 197–216.

Jansen, D. M., J. J. Stoorvogel, and R. A. Schipper. 1995. "Using Sustainability Indicators in Agricultural Land Use Analysis: An Example from Costa Rica." *Netherlands Journal of Agricultural Science*, 43: 61–82.

John, DeWitt. 1994. *Civic Environmentalism: Alternatives to Regulation in States and Communities*. Washington, DC: Congressional Quarterly Press.

Johnson, Kirk. 1993. *Beyond Polarization: Emerging Strategies for Reconciling Community and Environment*. Seattle: Northwest Policy Center, University of Washington.

Johnson, Kirk. 1995a. An Evaluation of the Henry's Fork Watershed Council. Seattle: Northwest Policy Center, University of Washington (October 9).

Johnson, Kirk. 1995b. "Watershed Council Helping to Rebuild Community in the Henry's Fork." *Community and the Environment*. A supplement to the Northwest Policy Center's *The Changing Northwest*. Seattle: Northwest Policy Center, University of Washington (December): 2–3.

Johnson, Kirk. 1996. *1996 Evaluation of the Henry's Fork Watershed Council*. Seattle: Northwest Policy Center, University of Washington (November 25).

Johnson, Kirk. 1997. *Toward a Sustainable Region: Evolving Strategies for Reconciling Community and Environment*. Seattle: Northwest Policy Center, University of Washington.

Jones, Lisa. 1996. "Howdy, Neighbor!: As a Last Resort, Westerners Start Talking to Each Other." *High Country News*, 28 (9) (May 13).

Kates, Robert W., William C. Clark, Robert Corell, J. Michael Hall, Carlo C. Jaeger, Ian Lowe, James J. McCarthy, Hans Joachim Schnellhuber, Bert Bolin, Nancy M. Dickson, Sylvie Faucheux, Gilberto C. Gallopin, Arnulf Grubler, Brian Huntley, Jill Jager, Narpat S. Jodha, Roger E. Kasperson, Akin Mabogunje, Pamela Matson, Harold Mooney, Berrien Moore III, Timothy O'Riordan, and Uno Svedin. 2001. "[Policy Forum: Environment and Development] Sustainability Science." *Science*, 292 (April 27): 641–642.

Katzmann, Robert. 1980. "Federal Trade Commission." In James Q. Wilson, ed., *The Politics of Regulation*, 152–187. New York: Basic Books.

Kaufman, Herbert. 1967. *The Forest Ranger*. Washington, DC: Resources for the Future Press.

Kearns, Kevin P. 1996. *Managing for Accountability: Preserving the Public Trust in Public and Nonprofit Organizations*. San Francisco: Jossey-Bass.

Kemmis, Daniel. 1990. *Community and the Politics of Place*. Norman: University of Oklahoma Press.

Kemmis, Daniel. 1999. Keynote Address: Eastern Idaho Watershed Conference, Pocatello (October 20).

Kemmis, Daniel. 2001. *This Sovereign Land: A New Vision for Governing the West*. Washington, DC: Island Press.

Kenney, Doug S. 1999. "The Historical and Sociopolitical Context of the Western Watersheds Movement." *Journal of the American Water Resources Association*, 35 (3): 493–503.

Kenney, Doug S. 2000. *Arguing about Consensus: Examining the Case against Western Watershed Initiatives and Other Collaborative Groups Active in Natural Resources Management*. Boulder: School of Law, University of Colorado.

Keohane, Robert O., and Elinor Ostrom. 1995. *Local Commons and Global Interdependence: Heterogeneity and Cooperation in Two Domains*. London: SAGE.

Keown, Larry D. 1998. *A Review of Techniques and Tools Used in Open Democratic Decision-Making*. Report prepared for the USDA Forest Service, Office of Cooperative Forestry and Ecosystem Management Coordination Staffs (January 20) Sheridan, WY.

Kettl, Donald F. 1983. *The Regulation of American Federalism*. Baltimore: Johns Hopkins University Press.

Kettl, Donald F. 1996. "Governing at the Millenium." In James Perry, ed., *The Handbook of Public Administration*, 5–18. San Francisco: Jossey-Bass.

Kettl, Donald F. 2000. "Reform, American Style." *Governing* (June): 14.

Khademian, Anne M. 1992. *The SEC and Capital Market Regulation: The Politics of Expertise*. Pittsburgh: University of Pittsburgh Press.

Khademian, Anne M. 1996. *Checking on Banks: Autonomy and Accountability in Three Federal Agencies*. Washington, DC: Brookings Institution.

King, Cheryl, and Camilla Stivers. 1998. *Government Is Us*. Thousand Oaks, CA: SAGE.

King, Gary, Robert O. Keohane, and Sidney Verba. 1994. *Designing Social Inquiry*. Princeton, NJ: Princeton University Press.

Klinkenborg, Verlyn. 1995. "A Group of Ranchers Have Banded Together to Protect the Range." *Audubon* (September-October): 35–45.

Klyza, Christopher M. 1996. *Who Controls Public Lands? Mining, Forestry, and Grazing Policies, 1870–1990*. Chapel Hill: University of North Carolina Press.

Knopman, Debra S. 1996. *Second Generation: A New Strategy for Environmental Protection*. Washington, DC: Center for Innovation and the Environment, The Progressive Foundation.

Knopman, Debra S., Megan M. Susman, and Marc K. Landy. 1999. "Civic Environmentalism: Tackling Tough Land-Use Problems with Innovative Governance." *Environment*, 41: 24–32.

Knott, Jack H., and Gary J. Miller. 1987. *Reforming Bureaucracy*. Englewood Cliffs, NJ: Prentice-Hall.

Kraft, Michael E., and Denise Scheberle. 1998. "Environmental Federalism at Decade's End: New Approaches and Strategies." *Publius: The Journal of Federalism*, 28 (1) (Winter): 131–146.

Kuklinski, James H., Paul J. Quirk, Jennifer Jerit, David Schwieder, and Robert F. Rich. 2000. "Misinformation and the Currency of Democratic Citizenship." *Journal of Politics*, 62 (3) (August): 790–816.

Lackey, Robert T. 1998. "Radically Contested Assertions in Ecosystem Management." Presented at the VII International Congress of Ecology, Florence, Italy (July 19–25).

Landy, Marc K. 1993. "Public Policy and Citizenship." In Helen Ingram and Steven Rathgeb Smith, eds., *Public Policy for Democracy*, 19–44. Washington, DC: Brookings Institution.

Landy, Marc K., Megan M. Susman, and Debra S. Knopman. 1999. *Civic Environmentalism in Action: A Field Guide to Regional and Local Initiatives*. Washington, DC: Progressive Policy Institute.

Leach, William D., and Pelkey, Neil W. 2000. *Making Watershed Partnerships Work: A Review of the Empirical Literature*. Unpublished manuscript, University of California at Davis.

Lead Partnership Group. 1996. *Proceedings of the Lead Partnership Group: Northern California/Southern Oregon Roundtable on Communities of Place, Partnerships, and Forest Health*. Blairsden, CA: Lead Partnership Group (October).

Lee, Kai N. 1993. *Compass and Gyroscope: Integrating Science and Politics for the Environment*. Washington, DC: Island Press.

Lehmbruch, Gerhard, and Philippe C. Schmitter, eds. 1982. *Patterns of Corporatist Policy-Making*. London: Sage Publications.

Levi, Margaret. 1998. "A State of Trust." In V. Braithwaite and M. Levi, eds., *Trust and Governance*, 77–101. New York: Russell Sage Foundation.

Light, Paul. 1993. *Monitoring Government: Inspectors General and the Search for Accountability*. Washington, DC: Brookings Institution.

Light, Paul. 1995. *Thickening Government: Federal Hierarchy and the Diffusion of Accountability*. Washington, DC: Brookings Institution.

Light, Paul. 1997. *The Tides of Reform*. New Haven, CT: Yale University Press.

Liroff, Richard A. 1976. *NEPA and Its Aftermath: The Formation of a National Policy for the Environment*. Bloomington: Indiana University Press.

Little, Jane Braxton. 1997. "The Feather River Alliance: Restoring Creeks and Communities in the Sierra Nevada." *Chronicle of Community*, 2 (1) (Autumn): 5–14.

Little, Jane Braxton. 1998. "A Quiet Victory in Quincy." *High Country News*, 30 (November 9): 2.

Locke, Michelle. 2001. "Western Drought Victims War over Water." *Washington Post* (July 8): A10.

Long, Norton E. [1947]. 1986. "Power and Administration." Reprinted in F. E. Rourke, ed., *Bureaucratic Power in National Policy Making*, 4th ed. Boston: Little, Brown.

Lowe, J. 1993. Memorandum to Forest Supervisors. USDA Forest Service, Pacific Northwest Region, Portland, OR (August 16).

Lowi, Theodore M., Jr. 1979. *The End of Liberalism*. Rev. ed. New York: Norton.

Lubell, Mark. 2000. "Attitudinal Support for Environmental Governance: Do Institutions Matter?" Paper presented at the Annual Meeting of the Midwest Political Science Association, Chicago, IL (April 27–30).

Lubell, Mark, Mark Schneider, John T. Scholz, and Mihriye Mete. 2002. "Watershed Partnerships and the Emergence of Collective Action Institutions." *American Journal of Political Science*, 46 (1) (January): 148–163.

Maass, Arthur. 1951. *Muddy Waters: The Army Engineers and Our Nation's Rivers*. Cambridge, MA: Harvard University Press.

MacDonnell, Lawrence J. 1999. *From Reclamation to Sustainability: Water, Agriculture, and the Environment in the American West*. Boulder: University Press of Colorado.

MacManus, Reed. 1997. "Fire Sale in the Forests: A Feel-Good Timber Plan That's Bad for Public Lands." *Sierra*, 82 (6) (November-December): 30–31.

Majone, Giandomenico. 1990. "Policy Analysis and Public Deliberation." In Robert Reich, ed., *The Power of Public Ideas*, 157–178. Cambridge, MA: Harvard University Press.

Malpai Borderlands Group. 1995. *1995–1997 Operating Plan*. Douglas, AZ. Malpai Borderlands Group.

Mann, Charles C., and Mark L. Plummer. 1995. *Noah's Choice: The Future of Endangered Species*. New York: Knopf.

Manning, Richard. 1997. "Working the Watershed." *High Country News* (March 17): 1, 8–12.

Mansbridge, Jane. 1980. *Beyond Adversary Democracy*. Chicago: University of Chicago Press.

Mansbridge, Jane. 1990. "The Rise and Fall of Self-Interest in the Explanation of Political Life." In Jane Mansbridge, ed., *Beyond Self-Interest*, Chicago: University of Chicago Press.

March, James, and Johan Olsen. 1989. *Rediscovering Institutions: The Organizational Basis of Politics*. New York: Free Press.

Marston, Ed. 1997a. "A Newsman's Overview of Willapa." *High Country News*, 29 (5) (March 17): 7.

Marston, Ed. 1997b. "The Timber Wars Evolve into a Divisive Attempt at Peace." *High Country News*, 29 (18) (September 29): 1, 8.

Marston, Ed. 1997c. "We're Much Stronger Together." *High Country News,* 29 (18) (September 29): 13.

Matthews, David. 1996. *Is There a Public for Public Schools?* Cleveland, OH: Kettering Foundation.

Mathews, David. 1999. "Megachallenges." *Higher Education Exchange* 78–88. Dayton, Ohio: The Kettering Foundation.

Mazmanian, Daniel A., and Michael E. Kraft, 1999a. "The Three Epochs of the Environmental Movement." In Daniel A. Mazmanian and Michael E. Kraft, eds., *Toward Sustainable Communities: Transition and Transformations in Environmental Policy,* 3–42. Cambridge, MA: MIT Press.

Mazmanian, Daniel A., and Michael E. Kraft, eds. 1999b. *Toward Sustainable Communities: Transition and Transformations in Environmental Policy.* Cambridge, MA: MIT Press.

Mazza, Patrick. 1997. "Cooptation or Constructive Engagement? Quincy Library Group's Effort to Bring Together Loggers and Environmentalists Under Fire." *Cascadia Planet.* Online at www.tnews.com/text/quincy_library.html (August 9); retrieved on October 31, 1997.

McCloskey, Michael. 1996. "The Skeptic: Collaboration Has Its Limits." *High Country News,* 28 (May 13). Online at www.hcn.org/1996/may13/dir/.html; retrieved on March 13, 1997.

McConnell, Grant. 1966. *Private Power and American Democracy.* New York: Knopf.

McGinnis, Michael V. 1999a. "Making the Watershed Connection." *Policy Studies Journal,* 27 (3): 497–501.

McGinnis, Michael V., ed. 1999b. *Bioregionalism.* New York: Routledge.

McKean, Margaret. 1992. "Success on the Commons: A Comparative Examination of Institutions for Common Property Resource Management." *Journal of Theoretical Politics,* 4: 247–281.

McKinney, Matthew J. 2001. "What Do We Mean by Consensus? Some Defining Principles." In Philip Brick, Donald Snow, and Sarah van de Wetering, eds., *Across the Great Divide: Explorations in Collaborative Conservation and the American West,* 33–40. Washington DC: Island Press.

Milward, H. Brinton. 1996. "The Changing Character of the Public Sector." In James Perry, ed., *The Handbook of Public Administration,* 77–91. San Francisco: Jossey-Bass.

Milward, H. Brinton, and Keith G. Provan. 1995. "A Preliminary Theory of Interorganizational Network Effectiveness: A Comparative Study of Four Community Mental Health Systems." *Administrative Science Quarterly,* 40: 1–33.

Milward, H. Brinton, and Keith Provan. 1999. *How Networks Are Governed.* Unpublished manuscript, Tucson, AZ: University of Arizona.

Mitro, M. 1999. *Sampling and Analysis Techniques and Their Application for Estimating Recruitment of Juvenile Rainbow Trout in the Henry's Fork of the Snake River, Idaho.* Doctoral dissertation, Montana State University, Bozeman.

Moe, Ronald C. 1994. "The Reinventing Government Exercise: Misinterpreting the Problem, Misjudging the Consequences." *Public Administration Review, 54* (2) (March-April): 111–122.

Moe, Ronald C., and Robert S. Gilmour 1995. "Rediscovering Principles of Administration: The Neglected Foundation of Public Law." *Public Administration Review, 55* (2): 135–146.

Moe, Terry M. 1989. The Politics of Bureaucratic Structure. In J. E. Chubb and E. Peterson, eds. *Can the Government Govern?* 267–329. Washington, DC: The Brookings Institution.

Morley, Elaine, Scott P. Bryant, and Harry P. Hatry. 2001. *Comparative Performance Measurement.* Washington, DC: Urban Institute Press.

Morone, James A. 1998. *The Democratic Wish: Popular Participation and Limits of American Government.* Rev. Ed. New Haven, CT: Yale University Press.

Morrell, Michael E. 1999. "Citizens' Evaluations of Participatory Democratic Procedures: Normative Theory Meets Empirical Science." *Political Research Quarterly, 52* (2) (June): 293–322.

Moscow-Pullman Daily News. 1999. "Rainbows Killed to Save Cutthroats." (October 11): B8.

Moseley, Cassandra. 1999. *New Ideas, Old Institutions: Environment, Community, and State in the Pacific Northwest.* Unpublished doctoral dissertation, Yale University.

Moseley, Cassandra. 2000. "Community Participation and Institutional Change: The Applegate Partnership and Federal Land Management in Southwest Oregon." Paper presented at the Western Political Science Association Annual Meeting, San Jose, CA (March 24–26).

Mosher, Frederick. 1982. *Democracy and the Public Service.* 2nd ed. New York: Oxford University Press.

Naess, Arne. 1983. "The Shallow and the Deep, Long-Range Ecology Movement: A Summary." *Inquiry, 16*: 95–100.

Natural Resources Law Center. 1996. *The Watershed Source Book: Watershed-Based Solutions to Natural Resource Problems.* Boulder: Natural Resources Law Center, University of Colorado.

Nelson, Robert H. 1995. *Public Lands and Private Rights: The Failure of Scientific Management.* Lanham, MD: Rowman and Littlefield.

Nelson, Robert H., and Mary M. Chapman. 1995. "Voices from the Heartland: New Tools for Decentralizing Management of the West's Public Lands." *Points West Special Report.* Denver: Center for the New West (December).

Newland, Chester A. 1996. "The National Government in Transition." In James Perry, ed., *The Handbook of Public Administration,* 19–35. San Francisco: Jossey-Bass.

Nijkamp, Peter, and Adriaan Perrels. 1994. *Sustainable Cities in Europe: A Comparative Analysis of Urban Energy-Environmental Policies.* London: Earthscan.

North, Douglass C. 1990. *Institutions, Institutional Change, and Economic Performance.* Cambridge: Cambridge University Press.

O'Leary, Rosemary, Robert F. Durant, Daniel J. Fiorino, and Paul S. Weiland. 1999. *Managing for the Environment: Understanding the Legal, Organizational, and Policy Challenges.* San Francisco: Jossey-Bass.

Osborne, David, and Ted Gaebler. 1993. *Reinventing Government.* New York: Penguin Books.

Ostrom, Elinor. 1990. *Governing the Commons: The Evolution of Institutions for Collective Action.* Cambridge: Cambridge University Press.

Ostrom, Elinor, and Edella Schlager. 1997. "The Formation of Property Rights." In Susan Hanna, Carl Folke, and Karl-Goran Maler, eds., *Rights to Nature: Ecological, Economic, Cultural, and Political Principles of Institutions for the Environment.* 127–156. Washington, DC: Island Press.

Paehlke, Robert. 1995. Environmental Values for a Sustainable Society: The Democratic Challenge. In Frank Fischer and Michael Black, eds., *Greening Environmental Policy: The Politics of a Sustainable Future,* 129–144. New York: St. Martin's Press.

Peck, F. Scott. 1978. *The Road Less Traveled.* New York: Simon and Schuster.

Pelkey, Neil, William Leach, Sky Harrison, Elizabeth Cook, Matthew Zafonte, and Paul Sabatier. 1999. "The Impacts of Social and Ecological Conditions on the Likelihood of Stakeholder-Based Resource Management Plans." Paper presented at the Annual Meeting of the Association for Public Policy Analysis and Management, Washington, DC (November 4–6).

Peters, B. Guy, and Donald J. Savoie. 1996. "Managing Incoherence: The Coordination and Empowerment Conundrum." *Public Administration Review,* 56 (May-June): 281–290.

Piore, Michael J. 1995. *Beyond Individualism.* Cambridge, MA: Harvard University Press.

Powell, Frona M. 1997. "Defining Harm under the Endangered Species Act: Implications of *Babbitt v. Sweet Home.*" *American Business Law Journal,* 33: 131–152.

Press, Daniel. 1994. *Democratic Dilemmas in the Age of Ecology: Trees and Toxics in the American West.* Durham, NC: Duke University Press.

Priester, Kevin. 1994. *Words into Action: A Community Assessment of the Applegate Valley.* Ashland, OR: The Rogue Institute of Ecology and Economy (May).

Priester, Kevin, and James Kent. 1997. "Social Ecology: New Pathways to Ecosystem Restoration." In J. E. Williams, M. P. Dombeck, and C. A. Wood, eds., *Watershed Restoration: Principles and Practices.* Bethesda, MD: American Fisheries Society.

Priester, Kevin, and Cassandra Moseley. 1997. *Applegate Valley Strategic Plan.* Ashland, OR: Rogue Institute for Ecology and Economy (July).

Provan, Keith, and H. Brinton Milward. 1995. "A Preliminary Theory of Network Effectiveness: A Comparative Study of Four Mental Health Systems." *Administrative Science Quarterly,* 40 (1) (March): 1–33.

Prugh, Thomas, Robert Costanza, and Herman Daly. 2000. *The Local Politics of Global Sustainability.* Washington, DC: Island Press.

Putnam, Robert D. 1993. *Making Democracy Work: Civic Traditions in Modern Italy.* Princeton, NJ: Princeton University Press.

Rabe, Barry G. 1986. *Fragmentation and Integration in State Environmental Management.* Washington, DC: Conservation Foundation.

Rabe, Barry G. 1991. "Impediments to Environmental Dispute Resolution in the American Political Context." In Miriam K. Mills, ed., *Alternative Dispute Resolution in the Public Sector,* 143–163. Chicago: Nelson-Hall.

Rabe, Barry G. 1994. *Beyond NIMBY: Hazardous Waste Siting in Canada and the United States.* Washington, DC: Brookings Institution.

Radin, Beryl A., Robert Agranoff, Ann O'M. Bowman, C. Gregory Buntz, J. Steven Ott, Barbara S. Romzek, and Robert H. Wilson. 1996. *New Governance for Rural America: Creating Intergovernmental Partnerships.* Lawrence: University of Kansas Press.

Radin, Beryl A., and Barbara S. Romzek. 1996. "Accountability Expectations in an Intergovernmental Arena." *Publius,* 26 (2): 59–81.

Reich, Robert. 1990a. "Policy Making in a Democracy." In Robert Reich, ed., *The Power of Public Ideas,* 123–156. Cambridge, MA: Harvard University Press.

Reich, Robert, ed. 1990b. *The Power of Public Ideas.* Cambridge, MA: Harvard University Press.

Reid, Rebecca, Linda Young, and Barry Russell. 1996. *Analysis of Demographic and Economic Aspects of the Applegate Watershed.* Ashland: Southern Oregon State University (September).

Reisner, Marc. 1986. *Cadillac Desert.* New York: Penguin Books.

Riebsame, William E., and James J. Robb. 1997. *Atlas of the New West: Portrait of a Changing Region.* New York: Norton.

Rieke, Betsy, and Doug Kenney. 1997. *Resource Management at the Watershed Level.* Report to the Western Water Policy Review Advisory Commission. Boulder: Natural Resources Law Center, University of Colorado.

Rogue Institute for Ecology and Economy. 1993. "Proposed Criteria and Indicators for Measuring Sustainable Forest Management in the U.S." Draft of Presentation at Applegate Partnership Meeting (September 10). Ashland, OR: Rogue Institute for Ecology and Economy.

Rolle, Su. 1997a. "The Applegate Partnership." In G. K. Meffe and C. R. Carroll, eds., *Principles in Conservation Biology,* 612–616. Sunderland, MA: Sinauer Associates.

Rolle, Su. 1997b. "5 Years and Still Cookin." Jacksonville, OR: Applegate Partnership.

Romzek, Barbara S. 1996. "Enhancing Accountability." In James Perry, ed., *The Handbook of Public Administration*, 97–114. San Francisco: Jossey-Bass.

Romzek, Barbara S., and Melvin J. Dubnick. 1994. "Issues of Accountability in Flexible Personnel Systems." In Patricia W. Ingraham and Barbara S. Romzek, eds., *New Paradigms for Government: Issues for the Changing Public Service*, 114–127. San Francisco: Jossey-Bass.

Rosenbaum, Walter A. 1976. "The Paradoxes of Public Participation." *Administration and Society*, 8 (November): 335–383.

Sandel, Michael J. 1996. *Democracy's Discontent: America in Search of a Public Philosophy.* Cambridge, MA: Belknap Press of Harvard University Press.

San Francisco Chronicle. 1997. "Who Should Determine the Fate of a Forest [editorial]?" (June 15): A22.

Sawicki, David, and Patrice Flynn. 1996. "Neighborhood Indicators: A Review of the Literature and Assessment of Conceptual and Methodological Issues." *Journal of the American Planning Association*, 62 (2): 165–183.

Scholz, John. 1991. "Cooperative Regulatory Enforcement and the Politics of Administrative Effectiveness." *American Political Science Review*, 85: 115–136.

Scott, James C. 1998. *Seeing Like a State: How Certain Schemes to Improve the Human Condition Have Failed.* New Haven, CT: Yale University Press.

Scruggs, Patricia, and Associates. 1996. *Colorado Forum: National and Community Indicators.* Proceedings of the Colorado Trust, Redefining Progress, and the White House Inter-Agency Working Group on Sustainable Development Indicators. Online at www.rprogress.org/events/cip 961122/cip_961122.html; retrieved on March 20, 2000.

Selznick, Philip. 1949. *TVA and the Grass Roots: A Study in the Sociology of Formal Organization.* Berkeley: University of California Press.

Shannon, Margaret, Victoria Sturtevant, and Dave Trask. 1997. *Organizing for Innovation: A Look at the Agencies and Organizations Responsible for Adaptive Management Areas: The Case of the Applegate AMA.* Report submitted to the Interagency Liaison for the U.S. Forest Service and Bureau of Land Management, Applegate Adaptive Management Area, Oregon (February).

Shea, Ruth E., and R.C. Drewien. 1999. *Evaluation of Efforts to Redistribute the Rocky Mountain Population of Trumpeter Swans 1986–1997.* Pocatello: Department of Biological Sciences, Idaho State University.

Shipley, Jack. 1996. "The Applegate Partnership." Jacksonville, OR. Online at www.watershed.org/news/sum%2095/applegate; retrieved on December 20, 1996.

Shipley, Jack, and Susan Shipley. 2000. Letter to the Editor. *The Applegator*, 5 (4) (January/February): 9.

Sierra Business Council. 1999. *The Sierra Nevada Wealth Index.* Sacramento, CA: Sierra Business Council.

Sirianni, Carmen, and Lewis Friedland. 1995a. "Civic Environmentalism." Online at www.cpn.org/sections/topics/envirspectives/civic_environmentalism.html; retrieved on October 15, 1997.

Sirianni, Carmen, and Lewis Friedland. 1995b. "Social Capital and Civic Innovation: Learning and Capacity Building from the 1960s to the 1990s." Paper presented at the annual American Sociological Association Conference, Washington, DC. (August 20).

Sirianni, Carmen, and Lewis Friedland. 2001. *Civic Innovation in America: Community Empowerment, Public Policy, and the Movement for Civic Renewal.* Berkeley: University of California Press.

Skocpol, Theda, and Morris P. Fiorina, eds. 1999. *Civic Engagement in American Democracy.* Washington, DC: Brookings Institution, and New York: Russell Sage Foundation.

Skowronek, Stephen. 1982. *Building a New American State.* Cambridge: Cambridge University Press.

Smith, Steven Rathgeb. 1993. "The New Politics of Contracting: Citizenship and the Nonprofit Role." In Helen Ingram and Steven Rathgeb Smith, eds., *Public Policy for Democracy,* 198–221. Washington, DC: Brookings Institution.

Snow, Donald. 1996. "Coming Home." *Chronicle of Community,* 1 (1) (Autumn): 40–43.

Snow, Donald. 1997. "Empire or Homelands? A Revival of Jeffersonian Democracy in the American West." In John A. Baden and Donald Snow, eds., *The Next West: Public Lands, Community, and Economy in the American West,* 181–203. Washington, DC: Island Press.

Snyder, Gary. 1990. *The Practice of the Wild.* New York: North Point Press.

Sparrow, Malcolm K. 1994. *Imposing Duties: Government's Changing Approach to Compliance.* Westport, CT: Praeger.

Sperry, Charles. 1997. *Community Development Groups: A Solution to Conflict in Western Montana.* Unpublished master's thesis, University of Montana, Missoula.

Stahl, Andy. 2001. "Ownership, Accountability, and Collaboration." In Philip Brick, Donald Snow, and Sarah van de Wetering, eds., *Across the Great Divide: Explorations in Collaborative Conservation and the American West,* 194–199. Washington, DC: Island Press.

Steel, Brent, and Edward Weber. 2001. "Environmental Policy, Devolution, and Public Opinion." *Global Environmental Change,* 11: 119–131.

Stegner, Wallace. 1969. *The Sound of Mountain Water.* New York: Doubleday.

Stone, Christopher. 1975. *Do Trees Have Standing? Toward Legal Rights for Natural Objects.* New York: Avon Press.

Sturtevant, Victoria E., and Jon I. Lange. 1995. *Applegate Partnership Case Study: Group Dynamics and Community Context.* Seattle: Pacific Northwest Research Station, U.S. Forest Service.

Susskind, Lawrence, and Jeffrey Cruickshank. 1987. *Breaking the Impasse: Consensual Approaches to Resolving Public Disputes.* New York: Basic Books.

Teton County Economic Development Council. 1997. *Welcome Home: A Homeowner's Handbook for Living in the Teton Valley.* Driggs, ID: Teton County Economic Development Council.

Thomas, Chant. 2000. Letter to the Editor. *The Applegator*, 5 (4) (January/February): 9.

Thompson, Frank J., and Norma M. Riccucci. 1998. "Reinventing Government." In Nelson Polsby, ed., *Annual Review of Political Science* 1, 231–257. Palo Alto, CA: Annual Reviews.

Tocqueville, Alexis de. [1835]. 1956. *Democracy in America.* New York: Times Mirror.

Torgerson, Douglas. 1998. "Limits of the Administrative Mind: The Problem of Defining Environmental Problems." In J. S. Dryzek and D. Schlosberg, eds., *Debating the Earth: The Environmental Politics Reader,* 110–127. New York: Oxford University Press.

Torgerson, Douglas. 1999. *The Promise of Green Politics: Environmentalism and the Public Sphere.* Durham, NC: Duke University Press.

U.S. Bureau of Land Management. 1998. *Ecosystem Restoration in the Ashland Resource Area.* Medford, OR: Department of the Interior.

U.S. Bureau of Land Management. 1999. Ecosystem Restoration Timber Sales (summary of timber sales from 1994 to 1998, work done with Applegate Partnership). Medford (Oregon) District, Ashland Resource Area. Medford, OR: U.S. Department of the Interior.

U.S. Bureau of Land Management and U.S. Forest Service. 1994. *Applegate Adaptive Management Area Ecosystem Health Assessment.* Medford, OR: Department of the Interior.

U.S. Bureau of Land Management and U.S. Forest Service. 1998. *Applegate Adaptive Management Area Guide.* Medford, OR: Department of the Interior.

U.S. Senate. 1997. "Workshop on Community-Based Approaches to Conflict Resolution in Public Land Management." Hearings before the Subcommittee on Forests and Public Land Management, Committee on Energy and Natural Resources, Washington, DC (May 22).

Van de Wetering, Sarah. 1996. "Ranchers as Land Stewards: Beyond Slogans." *The Chronicle of Community,* 1 (1) (Autumn): 26–27.

Van Kirk, Robert. 1996. "1996 Research Project Review." *Henry's Fork Foundation Quarterly Newsletter,* (Fall): 1, 4.

Van Kirk, Robert, and S. Beesley. 1999. *Downstream Migration of Rainbow Trout in the Buffalo River during 1997 and 1998.* Pocatello: Department of Biological Sciences, Idaho State University.

Van Kirk, Robert, and K. Giese. 1999. *Angler Effort and Catch on the Buffalo River during 1998 and Effects of the Buffalo River Fish Ladder on the Box Canyon Rainbow Trout Population.* Pocatello: Department of Biological Sciences, Idaho State University.

Van Kirk, Robert, and C. B. Griffin. 1997. "Building a Collaborative Process for Restoration: Henry's Fork of Idaho and Wyoming." In J. E. Williams, C. A. Wood, and M. P. Dombeck, eds., *Watershed Restoration: Principles and Practices,* 253–276. Bethesda, MD: American Fisheries Society.

Van Kirk, Robert, and R. Martin. 2000. "Interactions among Waterfowl Herbivory, Aquatic Vegetation, Fisheries and Flows below Island Park Dam." *Intermountain Journal of Sciences,* 6 (3) (September): 249–262.

Varma, Laurie. 1998. "Converging Paths: Researchers and Advocacy Groups Identify Clean Production Practices as a Preferred Route to Sustainability." *In-Sites: The Newsletter of the University of Tennessee's Waste Management Research and Education Institute,* 6 (3) (Fall): 2–3.

Wahl, Richard. 1989. *Markets for Federal Water.* Washington, DC: Resources for the Future.

Waldo, Dwight. 1984. *The Administrative State.* 2nd ed. New York: Holmes and Meier.

Walker, Doug. 1998. "Washington's Population Growth: A Big Challenge for Saving Our Ecosystems." *Washington Wildlands* (Winter): 12.

Walters, John. 1997. *Measuring Up: Governing's Guide for Performance Measurement.* Washington, DC: Governing Books.

Wapner, Paul. 1997. "Governance in Global Civil Society." In Oran R. Young, ed., *Global Governance: Drawing Insights from the Environmental Experience.* Cambridge, MA: MIT Press.

Warren, Mark. 1992. "Democratic Theory and Self Transformation." *American Political Science Review,* 86 (1) (March): 8–23.

Washington State Department of Fish and Wildlife. 2000. *Pacific Salmon and Wildlife.* Olympia: Washington State Department of Fish and Wildlife.

Weber, Edward P. 1998. *Pluralism by the Rules: Conflict and Cooperation in Environmental Regulation.* Washington, DC: Georgetown University Press.

Weber, Edward P. 1999a. "Changing Institutions and the Puzzle of Accountability: The Case of the Henry's Fork Watershed Council." Idaho National Engineering and Environmental Laboratory Working Paper Series, Idaho Falls (September).

Weber, Edward P. 1999b. *An Evaluation of Community Building for the Henry's Fork Watershed.* Report for the Brainerd Foundation, Seattle.

Weber, Edward P. 1999c. "The Question of Accountability in Historical Perspective: From Jackson to Contemporary Grass-Roots Ecosystem Management." *Administration and Society,* 31 (4) (September): 451–494.

Weber, Edward P. 2000a. "Cooperative Watershed Management and Research: The Case of the Henry's Fork Watershed Council." *Intermountain Journal of Sciences,* 6 (3): 293–311.

Weber, Edward P. 2000b. "A New Vanguard for the Environment: Grass-Roots Ecosystem Management as a New Environmental Movement." *Society and Natural Resources,* 13: 237–259.

Welch, Craig. 2001. "Both Sides Harden in Oregon Water Dispute." *Seattle Times* (July 9).

West, William F., and Joseph Cooper. 1989–90. "Legislative Influence v. Presi dential Dominance: Competing Models of Bureaucratic Control." *Political Science Quarterly,* 104(4): 581–606.

Western, David, and R. Michael Wright. 1994. *Natural Connections: Perspectives in Community-Based Conservation.* Washington, DC: Island Press.

Wilderness Society and National Audubon Society. 1996. *Salvage Logging in the National Forests: An Ecological, Economic, and Legal Assessment.* Washington, DC: Wilderness Society and National Audubon Society.

Wilhelm, Steve. 2000. "Exotic Invasion: Species Not Native to Region Are Damaging Northwest Waters." *Puget Sound Business Journal* (October 6). Online at www.bizjournals.com/seattle/stories/2000/10/09/focus1.html; retrieved on October 10, 2000.

Wilkinson, Charles F. 1992. *Crossing the Next Meridian: Land, Water, and the Future of the West.* Washington, DC: Island Press.

Willapa Alliance. 1995. *The Willapa Alliance: Completing the Foundation: 1995 Annual Report.* South Bend, WA: Willapa Alliance.

Willapa Alliance. 1996a. *The Bear River Watershed Restoration Partnership Project.* South Bend, WA: Willapa Alliance.

Willapa Alliance. 1996b. *Key Projects.* South Bend, WA: Willapa Alliance.

Willapa Alliance. 1996c. *Salmon Recovery Program.* South Bend, WA: Willapa Alliance.

Willapa Alliance. 1996d. *Science and Information Program.* South Bend, WA: Willapa Alliance.

Willapa Alliance. 1996e. *The Willapa Alliance: Willapa and the Web of Sustainability: Annual Report.* South Bend, WA: Willapa Alliance.

Willapa Alliance. 1996f. *Willapa Indicators for a Sustainable Community.* South Bend, WA: Willapa Alliance.

Willapa Alliance. 1996g. *Willapa Indicators for a Sustainable Community: 1996.* South Bend, WA: Willapa Alliance.

Willapa Alliance. 1996h. *The Willapa Sustainable Fisheries Harvest Program: Pilot Selective Hatchery Release and Harvest Project.* South Bend, WA: Willapa Alliance.

Willapa Alliance. 1997a. "The Willapa Alliance: Setting Course for Systemic Change: 1997 Annual Report [rough draft]." South Bend, WA: Willapa Alliance.

Willapa Alliance. 1997b. Willapa—*The Nature of Home: A Place-Based Community Education Program.* South Bend, WA: Willapa Alliance.

Willapa Alliance. 1998a. Cover letter to Willapa educators introducing the *Places, Faces and Systems of Home* School Resource Guide. South Bend, WA: Willapa Alliance.

Willapa Alliance. 1998b. *Places, Faces and Systems of Home: Connections to Local Learning.* South Bend, WA: Willapa Alliance.

Willapa Alliance. 1998c. *Willapa Indicators Community Conference* (advertisement, flyer). South Bend, WA: Willapa Alliance.

Willapa Alliance. 1998d. *Willapa Indicators for a Sustainable Community: 1998.* South Bend, WA: Willapa Alliance.

Willapa Alliance and Pacific County Economic Development Council. 1997. *Willapa Conflict Resolution Forums: Endangered Species, Habitat, Water and Natural Resource Users, and Regulations.* South Bend, WA: Willapa Alliance and Pacific County Economic Development Council.

Williams, Bruce A., and Albert R. Matheny. 1995. *Democracy, Dialogue, and Environmental Disputes: The Contested Languages of Social Regulation.* New Haven, CT: Yale University Press.

Wilson, James Q. 1985. "The Rediscovery of Character: Private Virtue and Public Policy." *Public Interest,* 81: 15–16.

Wilson, James Q. 1989. *Bureaucracy: What Government Agencies Do and Why They Do It.* New York: Basic Books.

Wilson, James Q. 1994. "Reinventing Public Administration." *Political Science and Politics,* 27 (4) (December): 667–673.

Wilson, Woodrow. [1887] 1997. "The Study of Administration." In J. M. Shafritz and A. C. Hyde, eds., *Classics of Public Administration,* 14–26. 4th ed. Fort Worth, TX: Harcourt Brace.

Wold, Amy. 1995a. " Fishermen, Groups Join Hands to Save Salmon." In *Willapa: Banking on the Bay,* compiled by the *Daily Astorian* and the *Chinook Observer:* 13.

Wold, Amy. 1995b. " Joint Venture Helps Streams Flow." In *Willapa: Banking on the Bay,* compiled by *The Daily Astorian* and *The Chinook Observer:* 14.

Wolf, Charles, Jr. 1993. *Markets or Government: Choosing Between Imperfect Alternatives.* 2nd ed. Cambridge, MA: MIT Press.

Wolf, Edward C. 1993. *A Tidewater Place: A Portrait of the Willapa Ecosystem.* South Bend, WA: Willapa Alliance.

Wolin, Sheldon. 1960. *Politics and Vision: Continuity and Innovation in Western Political Thought.* Boston: Little, Brown.

Wondollek, Julia M., and Steven L. Yaffee. 2000. *Making Collaboration Work: Lessons from Innovation in Natural Resource Management.* Washington, DC: Island Press.

World Commission on Environment and Development. 1987. *Our Common Future.* New York: Oxford University Press.

Yaffee, Steven L., Alli F. Phillips, Irene C. Frentz, Paul W. Hardy, Susanne M. Maleki, and Barbara E. Thorpe. 1996. *Ecosystem Management in the United States: An Assessment of Current Experience.* Washington, DC: Island Press.

Young, Oran R. 1989. *International Cooperation: Building Regimes for Natural Resources and the Environment.* Ithaca, NY: Cornell University Press.

Young, Oran R. 1994. *International Governance: Protecting the Environment in a Stateless Society.* Ithaca, NY: Cornell University Press.

Young, Oran R. 1996. "Rights, Rules, and Resources in International Society." In Susan Hanna, Carl Folke, and Karl-Goran Maler, eds., *Rights to Nature: Ecological, Economic, Cultural, and Political Principles of Institutions for the Environment,* 245–264. Washington, DC: Island Press.

Young, Oran R. 1997. "Rights, Rules, and Resources in World Affairs." In Oran R. Young, ed., *Global Governance: Drawing Insights from the Environmental Experience,* 1–23. Cambridge, MA: MIT Press.

Index